TECHNICAL MATH
DEMYSTIFIED

Demystified Series

Accounting Demystified
Advanced Statistics Demystified
Algebra Demystified
Anatomy Demystified
asp.net 2.0 Demystified
Astronomy Demystified
Biology Demystified
Biotechnology Demystified
Business Calculus Demystified
Business Math Demystified
Business Statistics Demystified
C++ Demystified
Calculus Demystified
Chemistry Demystified
College Algebra Demystified
Corporate Finance Demystified
Data Structures Demystified
Databases Demystified
Differential Equations Demystified
Digital Electronics Demystified
Earth Science Demystified
Electricity Demystified
Electronics Demystified
Environmental Science Demystified
Everyday Math Demystified
Forensics Demystified
Genetics Demystified
Geometry Demystified
Home Networking Demystified
Investing Demystified
Java Demystified
JavaScript Demystified
Linear Algebra Demystified
Macroeconomics Demystified

Management Accounting Demystified
Math Proofs Demystified
Math Word Problems Demystified
Medical Terminology Demystified
Meteorology Demystified
Microbiology Demystified
Microeconomics Demystified
Nanotechnology Demystified
OOP Demystified
Options Demystified
Organic Chemistry Demystified
Personal Computing Demystified
Pharmacology Demystified
Physics Demystified
Physiology Demystified
Pre-Algebra Demystified
Precalculus Demystified
Probability Demystified
Project Management Demystified
Psychology Demystified
Quality Management Demystified
Quantum Mechanics Demystified
Relativity Demystified
Robotics Demystified
Six Sigma Demystified
SQL Demystified
Statistics Demystified
Technical Math Demystified
Trigonometry Demystified
UML Demystified
Visual Basic 2005 Demystified
Visual C# 2005 Demystified
XML Demystified

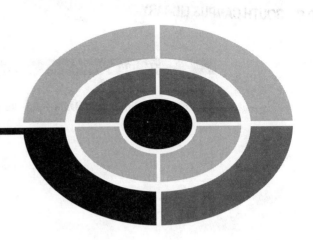

TECHNICAL MATH
DEMYSTIFIED

STAN GIBILISCO

McGRAW-HILL

New York   Chicago   San Francisco   Lisbon   London
Madrid   Mexico City   Milan   New Delhi   San Juan
Seoul   Singapore   Sydney   Toronto

The *McGraw·Hill* Companies

Cataloging-in-Publication Data is on file with the Library of Congress

1 2 3 4 5 6 7 8 9 0 DOC/DOC 0 1 2 1 0 9 8 7 6

ISBN 0-07-145949-9

The sponsoring editor for this book was Judy Bass and the production supervisor was Pamela A. Pelton. It was set in Times Roman by D&P Editorial Services. The art director for the cover was Margaret Webster-Shapiro; the cover designer was Handel Low.

Printed and bound by RR Donnelley.

This book is printed on acid-free paper.

McGraw-Hill books are available at special quantity discounts to use as premiums and sales promotions, or for use in corporate training programs. For more information, please write to the Director of Special Sales, McGraw-Hill Professional, Two Penn Plaza, New York, NY 10121-2298. Or contact your local bookstore.

To Samuel, Tim, and Tony

CONTENTS

Preface   xiii

Acknowledgments   xv

CHAPTER 1   **Numbering Systems**   1
Sets   1
Denumerable Number Sets   6
Bases 10, 2, 8, and 16   10
Nondenumerable Number Sets   15
Special Properties of Complex Numbers   20
Quick Practice   24
Quiz   27

CHAPTER 2   **Principles of Calculation**   29
Basic Principles   29
Miscellaneous Principles   33
Advanced Principles   37
Approximation and Precedence   42
Quick Practice   46
Quiz   47

CHAPTER 3   **Scientific Notation**   51
Powers of 10   51
Calculations in Scientific Notation   57
Significant Figures   61
Quick Practice   65
Quiz   67

CHAPTER 4   **Coordinates in Two Dimensions**   **71**
Cartesian Coordinates   71
Simple Cartesian Graphs   74
Polar Coordinates   80
Navigator's Coordinates   87
Coordinate Conversions   89
Other Coordinate Systems   92
Quick Practice   99
Quiz   101

CHAPTER 5   **Coordinates in Three Dimensions**   **105**
Cartesian 3-Space   105
Other 3D Coordinate Systems   108
Hyperspace   113
Quick Practice   119
Quiz   122

CHAPTER 6   **Equations in One Variable**   **125**
Operational Rules   125
Linear Equations   127
Quadratic Equations   130
Higher-Order Equations   134
Quick Practice   137
Quiz   139

CHAPTER 7   **Multivariable Equations**   **143**
2×2 Linear Equations   143
3×3 Linear Equations   148
2×2 General Equations   152
Graphic Solution of Pairs of Equations   154
Quick Practice   158
Quiz   160

CHAPTER 8   **Perimeter and Area in Two Dimensions**   **163**
Triangles   163
Quadrilaterals   166
Regular Polygons   171
Circles and Ellipses   172
Other Formulas   175
Quick Practice   180
Quiz   182

CHAPTER 9   **Surface Area and Volume in
            Three Dimensions**   **185**
Straight-Edged Objects   185
Cones and Cylinders   191
Other Solids   198
Quick Practice   202
Quiz   204

CHAPTER 10   **Boolean Algebra**   **207**
Operations, Relations, and Symbols   207
Truth Tables   212
Some Boolean Laws   216
Quick Practice   220
Quiz   223

CHAPTER 11   **Trigonometric Functions**   **227**
The Unit Circle   227
Primary Circular Functions   229
Secondary Circular Functions   232
The Right Triangle Model   234
Trigonometric Identities   237
Quick Practice   245
Quiz   248

CHAPTER 12   **Vectors in Two and Three Dimensions**   **251**
Vectors in the Cartesian Plane   251
Vectors in the Polar Plane   256
Vectors in Cartesian 3-Space   259
Standard Form of a Vector   264
Basic Properties   267
Other Properties   275
Quick Practice   278
Quiz   280

CHAPTER 13   **Logarithmic and Exponential Functions**   **283**
Logarithmic Functions   284
How Logarithmic Functions Behave   287
Exponential Functions   290
How Exponential Functions Behave   293
Quick Practice   298
Quiz   300

CHAPTER 14   **Differentiation in One Variable**   **305**
Definition of the Derivative   305
Properties of Derivatives   311
Properties of Curves   315
Derivatives of Wave Functions   323
Quick Practice   329
Quiz   331

CHAPTER 15   **Integration in One Variable**   **337**
What Is Integration?   337
Basic Properties of Integration   341
A Few More Formulas   343
Integrals of Wave Functions   348
Examples of Definite Integration   354
Quick Practice   358
Quiz   361

Final Exam   365

Answers to Quiz and Exam Questions   395

Suggested Additional Reading   399

Index   401

PREFACE

This book is written for people who want to refresh or improve their mathematical skills, especially in fields applicable to science and engineering. The course can be used for self-teaching without the aid of an instructor, but it can also be useful as a supplement in a classroom, tutored, or home-schooling environment. If you are changing careers, and your new work will involve more mathematics than you've been used to doing, this book should help you prepare.

If you want to get the most out of this book, you should have completed high-school algebra, high-school geometry and trigonometry, and a first-year course in calculus. You should be familiar with the concepts of rational, real, and complex numbers, linear equations, quadratic equations, the trigonometric functions, coordinate systems, and the differentiation and integration of functions in a single variable.

This book contains plenty of examples and practice problems. Each chapter ends with a multiple-choice quiz. There is a multiple-choice final exam at the end of the course. The questions in the quizzes and the exam are similar in format to the questions in standardized tests.

The chapter-ending quizzes are open-book. You may refer to the chapter texts when taking them. When you think you're ready, take the quiz, write down your answers, and then give your list of answers to a friend. Have the friend tell you your score, but not which questions you got wrong. The answers are listed in the back of the book. Stick with a chapter until you get most, and preferably all, of the quiz answers correct.

The final exam contains questions drawn uniformly from all the chapters. It is a closed-book test. Don't look back at the text when taking it. A satisfactory score is at least three-quarters of the answers correct (I suggest you shoot for 90 percent). With the final exam, as with the quizzes, have a friend tell you your score without letting you know which questions you missed. That way, you will

not subconsciously memorize the answers. You can check to see where your knowledge is strong and where it is weak.

I recommend that you complete one chapter a week. An hour or two daily ought to be enough time for this. When you're done with the course, you can use this book as a permanent reference.

Suggestions for future editions are welcome.

STAN GIBILISCO

ACKNOWLEDGMENTS

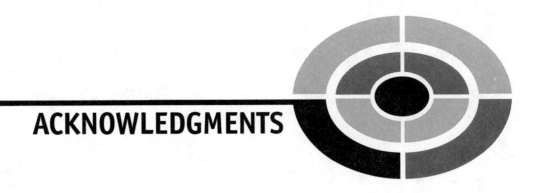

I extend thanks to my nephew Tony Boutelle, a student at Macalester College in St. Paul. He spent many hours helping me proofread the manuscript, and he offered insights and suggestions from the point of view of the intended audience.

TECHNICAL MATH
DEMYSTIFIED

Numbering Systems

This chapter covers the basic properties of sets and numbers. Familiarity with these concepts is important in order to gain a solid working knowledge of applied mathematics. For reference, and to help you navigate the notation you'll find in this book, Table 1-1 lists and defines the symbols commonly used in technical mathematics.

Sets

A *set* is a collection or group of definable *elements* or *members*. A set element can be anything—even another set. Some examples of set elements in applied mathematics and engineering are:

- Points on a line
- Instants in time

Table 1-1. Symbols commonly used in mathematics.

Symbol	Description	Symbol	Description
( )	Quantification; read "the quantity"	⊆	Subset; read "is a subset of"
[ ]	Quantification; used outside ( )	∈	Element; read "is an element of" or "is a member of"
{ }	Quantification; used outside [ ]		
{ }	Braces; objects between them are elements of a set	∉	Nonelement; read "is not an element of" or "is not a member of"
⇒	Logical implication or "if/then" operation; read "implies"	=	Equality; read "equals" or "is equal to"
		≠	Not-equality; read "does not equal" or "is not equal to"
⇔	Logical equivalence; read "if and only if"		
∀	Universal quantifier; read "For all" or "For every"	≈	Approximate equality; read "is approximately equal to"
∃	Existential quantifier; read "For some"	<	Inequality; read "is less than"
:	Logical expression; read "such that"	≤	Equality or inequality; read "is less than or equal to"
|	Logical expression; read "such that"	>	Inequality; read "is greater than"
&	Logical conjunction; read "and"	≥	Equality or inequality; read "is greater than or equal to"
∨	Logical disjunction; read "or"		
¬	Logical negation; read "not"	+	Addition; read "plus"
N	The set of natural numbers	−	Subtraction, read "minus"
Z	The set of integers	×	Multiplication; read "times" or "multiplied by"
Q	The set of rational numbers	*	Multiplication; read "times" or "multiplied by"
R	The set of real numbers	·	Multiplication; read "times" or "multiplied by"
ℵ	Transfinite (or infinite) cardinal number	÷	Quotient; read "over" or "divided by"
∅	The set with no elements; read "the empty set" or "the null set"	/	Quotient; read "over" or "divided by"
		!	Product of all natural numbers from 1 up to a certain value; read "factorial"
∩	Set intersection; read "intersect"		
∪	Set union; read "union"	×	Cross (vector) product of vectors; read "cross"
⊂	Proper subset; read "is a proper subset of"	•	Dot (scalar) product of vectors; read "dot"

- Coordinates in a plane.
- Coordinates in space.
- Points, lines, or curves on a graph.
- Digital logic states.
- Data bits, bytes, or characters.
- Subscribers to a network.
- Wind-velocity vectors at points in the eyewall of a hurricane.
- Force vectors at points along the length of a bridge.

If an element a is contained in a set A, then the fact is written like this:

$$a \in A$$

SET INTERSECTION

The *intersection* of two sets A and B, written $A \cap B$, is the set C consisting of the elements in both sets A and B. The following statement is valid for every element x:

$$x \in C \text{ if and only if } x \in A \text{ and } x \in B$$

SET UNION

The *union* of two sets A and B, written $A \cup B$, is the set C consisting of the elements in set A or set B (or both). The following statement is valid for every element x:

$$x \in C \text{ if and only if } x \in A \text{ or } x \in B$$

COINCIDENT SETS

Two nonempty sets A and B are *coincident* if and only if they are identical. That means that for all elements x, the following statements are both true:

$$\text{If } x \in A, \text{ then } x \in B$$

$$\text{If } x \in B, \text{ then } x \in A$$

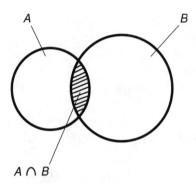

Fig. 1-1. The intersection of two non-disjoint, noncoincident sets A and B.

DISJOINT SETS

Two sets A and B are *disjoint* if and only if both sets contain at least one element, but there is no element that is in both sets. All three of the following conditions must be met:

$$A \neq \varnothing$$
$$B \neq \varnothing$$
$$A \cap B = \varnothing$$

where $\varnothing$ denotes the *empty set*, also called the *null set*.

VENN DIAGRAMS

The intersection and union of nonempty sets can be conveniently illustrated by *Venn diagrams*. Figure 1-1 is a Venn diagram that shows the intersection of two sets that are nondisjoint (they overlap) and noncoincident (they are not identical). Set $A \cap B$ is the cross-hatched area, common to both sets A and B. Figure 1-2 shows the union of the same two sets. Set $A \cup B$ is the shaded area, representing elements that are in set A or in set B, or both.

SUBSETS

A set A is a *subset* of a set B, written $A \subseteq B$, if and only if any element x in set A is also in set B. The following logical statement holds true for all elements x:

$$\text{If } x \in A, \text{ then } x \in B$$

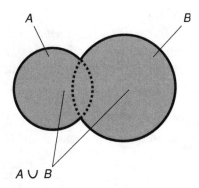

Fig. 1-2. The union of two non-disjoint, noncoincident sets A and B.

PROPER SUBSETS

A set A is a *proper subset* of a set B, written $A \subset B$, if and only if any element x in set A is in set B, but the two sets are not coincident. The following logical statements both hold true for all elements x:

$$\text{If } x \in A, \text{ then } x \in B$$

$$A \neq B$$

CARDINALITY

The *cardinality* of a set is the number of elements in the set. The null set has *zero cardinality*. The set of data bits in a digital image, stars in a galaxy, or atoms in a chemical sample has *finite cardinality*. Some number sets have *denumerably infinite cardinality*. Such a set can be fully defined by a listing scheme. An example is the set of all counting numbers $\{1, 2, 3,\dots\}$. Not all infinite sets are denumerable. There are some sets with *non-denumerably infinite cardinality*. This kind of set cannot be fully defined in terms of any listing scheme. An example is the set of all real numbers, which are those values that represent measurable physical quantities (and their negatives).

 PROBLEM 1-1
Find the union and the intersection of the following two sets:

$$S = \{2, 3, 4, 5, 6\}$$
$$T = \{4, 5, 6, 7, 8\}$$

SOLUTION 1-1

The union of the two sets is the set $S \cup T$ consisting of all the elements in one or both of the sets S and T. It is only necessary to list an element once if it happens to be in both sets. Thus:

$$S \cup T = \{2, 3, 4, 5, 6, 7, 8\}$$

The intersection of the two sets is the set $S \cap T$ consisting of all the elements that are in both of the sets S and T:

$$S \cap T = \{4, 5, 6\}$$

PROBLEM 1-2

In Problem 1-1, four sets are defined: S, T, $S \cup T$, and $S \cap T$. Are there any cases in which one of these sets is a proper subset of one or more of the others? If so, show any or all examples, and express these examples in mathematical symbology.

SOLUTION 1-2

Set S is a proper subset of $S \cup T$. Set T is also a proper subset of $S \cup T$. We can write these statements formally as follows:

$$S \subset (S \cup T)$$
$$T \subset (S \cup T)$$

It also turns out, in the situation of Problem 1-1, that the set $S \cap T$ is a proper subset of S, and the set $S \cap T$ is a proper subset of T. In formal symbology, these statements are:

$$(S \cap T) \subset S$$
$$(S \cap T) \subset T$$

The parentheses are included in these symbolized statements in order to prevent confusion as to how they are supposed to be read. A mathematical purist might point out that, in these examples, parentheses are not necessary, because the meanings of the statements are evident from their context alone.

Denumerable Number Sets

The set of familiar *natural numbers*, the set of *integers* (natural numbers and their negatives, including 0), and the set of *rational numbers* are examples of sets with denumerable cardinality. This means that they can each be arranged in the form of an infinite (open-ended) list in which each element can be assigned a counting number that defines its position in the list.

NATURAL NUMBERS

The natural numbers, also known as *whole numbers*, are built up from a starting point of 0. The set of natural numbers is denoted N, and is commonly expressed like this:

$$N = \{0, 1, 2, 3, \ldots, n, \ldots\}$$

In some texts, zero is not included, so the set of natural numbers is defined as follows:

$$N = \{1, 2, 3, 4, \ldots, n, \ldots\}$$

This second set, starting with 1 rather than 0, is sometimes called the set of *counting numbers*.

The natural numbers can be expressed as points along a horizontal half-line or ray, where quantity is directly proportional to displacement (Fig. 1-3). In the illustration, natural numbers correspond to points where hash marks cross the ray. Increasing numerical values correspond to increasing displacement toward the right. Sometimes the ray is oriented vertically, and increasing values correspond to displacement upward.

INTEGERS

The set of natural numbers can be duplicated and inverted to form an identical, mirror-image set:

$$-N = \{0, -1, -2, -3, \ldots, -n, \ldots\}$$

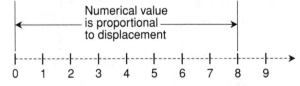

Fig. 1-3. The natural numbers can be depicted as discrete points on a half-line or ray. The numerical value is directly proportional to the displacement.

The union of this set with the set of natural numbers produces the set of integers, commonly denoted **Z**:

$$Z = N \cup -N = \{\ldots, -n, \ldots, -2, -1, 0, 1, 2, \ldots, n, \ldots\}$$

Integers can be expressed as points along a horizontal line, where positive quantity is directly proportional to displacement toward the right, and negative quantity is directly proportional to displacement toward the left (Fig. 1-4). In the illustration, integers correspond to points where hash marks cross the line. Sometimes a vertical line is used. In most such cases, positive values correspond to upward displacement, and negative values correspond to downward displacement. The set of natural numbers is a proper subset of the set of integers. Stated symbolically:

$$N \subset Z$$

RATIONAL NUMBERS

A rational number (the term derives from the word *ratio*) is a number that can be expressed as, or reduced to, the quotient of two integers, a and b, where b is positive. The standard form for a rational number r is:

$$r = a/b$$

The set of all possible quotients of this form composes the entire set of rational numbers, denoted **Q**. Thus, we can write:

$$Q = \{x \mid x = a/b, \text{ where } a \in Z, b \in Z, \text{ and } b > 0\}$$

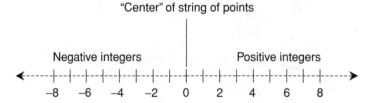

Fig. 1-4. The integers can be depicted as discrete points on a horizontal line. Displacement to the right corresponds to positive values, and displacement to the left corresponds to negative values.

The set of integers is a proper subset of the set of rational numbers. The natural numbers, the integers, and the rational numbers have the following relationship:

$$N \subset Z \subset Q$$

DECIMAL EXPANSIONS

Rational numbers can be denoted in *decimal form* as an integer followed by a period (*radix point*, also called a *decimal point*), and then followed by a sequence of digits. The digits to the right of the radix point always exist in either of two forms:

- A finite string of digits beyond which all digits are zero.
- An infinite string of digits that repeat in cycles.

Here are two examples of the first form, known as *terminating decimal numbers*:

$$3/4 = 0.750000\ldots$$
$$-9/8 = -1.1250000\ldots$$

Here are two examples of the second form, known as *nonterminating, repeating decimal numbers*:

$$1/3 = 0.33333\ldots$$
$$-123/999 = -0.123123123\ldots$$

PROBLEM 1-3
Of what use are negative numbers? How can you have a quantity smaller than zero? Isn't that like having less than none of something?

SOLUTION 1-3
Negative numbers are surprisingly common. Most people have experienced temperature readings that are "below zero," especially if the Celsius scale is used. Sometimes, driving in reverse instead of in forward gear is considered to be "negative velocity." Some people carry a "negative bank balance" for a short time. The government always seems to have a "deficit," and corporations often operate "in the red."

PROBLEM 1-4
Express the number 2457/9999 as a nonterminating, repeating decimal.

 SOLUTION 1-4
If you have a calculator that displays plenty of digits (the scientific-mode calculator in Windows XP is excellent), you can find this easily:

$$2457/9999 = 0.245724572457\ldots$$

The sequence of digits 2457 keeps repeating "forever." Note that this number is rational because it is the quotient of two integers, even though it is not a terminating decimal. That is, it can't be written out fully in decimal form using only a finite number of digits to the right of the radix point.

Bases 10, 2, 8, and 16

The numbering system used by people (as opposed to computers and calculators) in everyday life is the *decimal number system*, based on powers of 10. Machines, in contrast, generally perform calculations using numbering systems based on powers of 2.

DECIMAL NUMBERS

The *decimal number system* is also called *modulo 10*, *base 10*, or *radix 10*. Digits are elements of the set {0, 1, 2, 3, 4, 5, 6, 7, 8, 9}. The digit immediately to the left of the radix (decimal) point is multiplied by 10^0, or 1. The next digit to the left is multiplied by 10^1, or 10. The power of 10 increases as you move further to the left. The first digit to the right of the radix point is multiplied by a factor of 10^{-1}, or 1/10. The next digit to the right is multiplied by 10^{-2}, or 1/100. This continues as you go further to the right. Once the process of multiplying each digit is completed, the resulting values are added. This is what is represented when you write a decimal number. For example:

$$2704.53816 = (2 \times 10^3) + (7 \times 10^2) + (0 \times 10^1) + (4 \times 10^0)$$
$$+ (5 \times 10^{-1}) + (3 \times 10^{-2}) + (8 \times 10^{-3}) + (1 \times 10^{-4}) + (6 \times 10^{-5})$$

The parentheses are added for clarity.

BINARY NUMBERS

The *binary number system* is a method of expressing numbers using only the digits 0 and 1. It is sometimes called *modulo 2, base 2*, or *radix 2*. The digit immediately to the left of the radix point is the "ones" digit. The next digit to the left is a "twos" digit; after that comes the "fours" digit. Moving further to the left, the digits represent 8, 16, 32, 64, and so on, doubling every time. To the right of the radix point, the value of each digit is cut in half again and again, that is, 1/2, 1/4, 1/8, 1/16, 1/32, 1/64, and so on.

Consider an example using the decimal number 94:

$$94 = (4 \times 10^0) + (9 \times 10^1)$$

In the binary number system the breakdown is:

$$1011110 = (0 \times 2^0) + (1 \times 2^1) + (1 \times 2^2)$$
$$+ (1 \times 2^3) + (1 \times 2^4) + (0 \times 2^5) + (1 \times 2^6)$$

When you work with a computer or calculator, you give it a decimal number that is converted into binary form. The computer or calculator does its operations with zeros and ones, which are represented by different voltages or signals in electronic circuits. When the process is complete, the machine converts the result back into decimal form for display.

OCTAL NUMBERS

Another numbering scheme, called the *octal number system*, has eight symbols, or 2 cubed (2^3). It is also called *modulo 8, base 8*, or *radix 8*. Every digit is an element of the set {0, 1, 2, 3, 4, 5, 6, 7}. Counting thus proceeds from 7 directly to 10, from 77 directly to 100, from 777 directly to 1000, and so on. There are no numerals 8 or 9. In octal notation, decimal 8 is expressed as 10, and decimal 9 is expressed as 11.

HEXADECIMAL NUMBERS

Yet another scheme, commonly used in computer practice, is the *hexadecimal number system*, so named because it has 16 symbols, or 2 to the fourth power (2^4). These digits are the usual 0 through 9 plus six more, represented by A through F, the first six letters of the alphabet. The digit set is {0, 1, 2, 3, 4, 5, 6, 7, 8, 9, A, B, C, D, E, F}. In this number system, A is the equivalent of decimal 10, B is the equivalent of decimal 11, C is the equivalent of decimal 12, D is the equivalent of decimal 13, E is the equivalent of decimal 14, and F is the equivalent of decimal 15. This system is also called *modulo 16, base 16,* or *radix 16.*

COMPARISON OF VALUES

In Table 1-2, numerical values are compared in modulo 10 (decimal), 2 (binary), 8 (octal), and 16 (hexadecimal), for the decimal numbers 0 through 64. In general, as the *modulus* (or number base) increases, the numeral representing a given value becomes "smaller."

PROBLEM 1-5
Express the binary number 10011011 in decimal form.

SOLUTION 1-5
Working from right to left, the digits add up as follows:

$$10011011 = (1 \times 2^0) + (1 \times 2^1) + (0 \times 2^2) + (1 \times 2^3)$$
$$+ (1 \times 2^4) + (0 \times 2^5) + (0 \times 2^6) + (1 \times 2^7)$$
$$= (1 \times 1) + (1 \times 2) + (0 \times 4) + (1 \times 8)$$
$$+ (1 \times 16) + (0 \times 32) + (0 \times 64) + (1 \times 128)$$
$$= 1 + 2 + 0 + 8 + 16 + 0 + 0 + 128$$
$$= 155$$

PROBLEM 1-6
Express the decimal number 1,000,000 in hexadecimal form.

SOLUTION 1-6
Solving a problem like this is straightforward, but the steps are tricky, tedious, and repetitive. Some calculators will perform conversions like this directly, but if you don't have access to one, you can proceed in the following manner.

Table 1-2. Comparison of numerical values for decimal numbers 0 through 64.

Decimal	Binary	Octal	Hexadecimal	Decimal	Binary	Octal	Hexadecimal
0	0	0	0	33	100001	40	21
1	1	1	1	34	100010	42	22
2	10	2	2	35	100011	43	23
3	11	3	3	36	100100	44	24
4	100	4	4	37	100101	45	25
5	101	5	5	38	100110	46	26
6	110	6	6	39	100111	47	27
7	111	7	7	40	101000	50	28
8	1000	10	8	41	101001	51	29
9	1001	11	9	42	101010	52	2A
10	1010	12	A	43	101011	53	2B
11	1011	13	B	44	101100	54	2C
12	1100	14	C	45	101101	55	2D
13	1101	15	D	46	101110	56	2E
14	1110	16	E	47	101111	57	2F
15	1111	17	F	48	110000	60	30
16	10000	20	10	49	110001	61	31
17	10001	21	11	50	110010	62	32
18	10010	22	12	51	110011	63	33
19	10011	23	13	52	110100	64	34
20	10100	24	14	53	110101	65	35
21	10101	25	15	54	110110	66	36
22	10110	26	16	55	110111	67	37
23	10111	27	17	56	111000	70	38
24	11000	30	18	57	111001	71	39
25	11001	31	19	58	111010	72	3A
26	11010	32	1A	59	111011	73	3B
27	11011	33	1B	60	111100	74	3C
28	11100	34	1C	61	111101	75	3D
29	11101	35	1D	62	111110	76	3E
30	11110	36	1E	63	111111	77	3F
31	11111	37	1F	64	1000000	100	40
32	100000	40	20				

The values of the digits in a whole (that is, nonfractional) hexadecimal number, proceeding from right to left, are natural-number powers of 16. That means a whole hexadecimal number n_{16} has this form:

$$n_{16} = \ldots + (f \times 16^5) + (e \times 16^4) + (d \times 16^3)$$
$$+ (c \times 16^2) + (b \times 16^1) + (a \times 16^0)$$

where $a, b, c, d, e, f, \ldots$ are single-digit hexadecimal numbers from the set {0, 1, 2, 3, 4, 5, 6, 7, 8, 9, A, B, C, D, E, F}.

In order to find the hexadecimal value of decimal 1,000,000, first find the largest power of 16 that is less than or equal to 1,000,000. This is $16^4 = 65,536$. Then, divide 1,000,000 by 65,536. This equals 15 and a remainder. The decimal 15 is represented by the hexadecimal F. We now know that the decimal number 1,000,000 looks like this in hexadecimal form:

$$(F \times 16^4) + (d \times 16^3) + (c \times 16^2) + (b \times 16^1) + a = Fdcba$$

In order to find the value of d, note that $15 \times 16^4 = 983,040$. This is 16,960 smaller than 1,000,000. That means we must find the hexadecimal equivalent of decimal 16,960 and add it to hexadecimal F0000. The largest power of 16 that is less than or equal to 16,960 is 16^3, or 4096. Divide 16,960 by 4096. This equals 4 and a remainder. We now know that $d = 4$ in the above expression, so decimal 1,000,000 is equivalent to the following in hexadecimal form:

$$(F \times 16^4) + (4 \times 16^3) + (c \times 16^2) + (b \times 16^1) + a = F4cba$$

In order to find the value of c, note that $(F \times 16^4) + (4 \times 16^3) = 983,040 + 16,384 = 999,424$. This is 576 smaller than 1,000,000. That means we must find the hexadecimal equivalent of decimal 576 and add it to hexadecimal F4000. The largest power of 16 that is less than or equal to 576 is 16^2, or 256. Divide 576 by 256. This equals 2 and a remainder. We now know that $c = 2$ in the above expression, so decimal 1,000,000 is equivalent to the following in hexadecimal form:

$$(F \times 16^4) + (4 \times 16^3) + (2 \times 16^2) + (b \times 16^1) + a = F42ba$$

In order to find the value of b, note that $(F \times 16^4) + (4 \times 16^3) + (2 \times 16^2) = 983,040 + 16,384 + 512 = 999,936$. This is 64 smaller than 1,000,000. That means we must find the hexadecimal equivalent of decimal 64 and add it to hexadecimal F4200. The largest power of 16 that is less than or equal to 64 is 16^1, or 16. Divide 64 by 16. This equals 4 without any remainder. We now know that $b = 4$ in the above expression, so decimal 1,000,000 is equivalent to hexadecimal:

$$(F \times 16^4) + (4 \times 16^3) + (2 \times 16^2) + (4 \times 16^1) + a = F424a$$

There was no remainder when we found b. Thus, all the digits to the right of b (in this case, that means only the digit a) are equal to 0. Decimal 1,000,000 is therefore equivalent to hexadecimal F4240.

Checking, note that the hexadecimal F4240 breaks down as follows when converted to decimal form, proceeding from right to left:

$$F4240 = (0 \times 16^0) + (4 \times 16^1) + (2 \times 16^2) + (4 \times 16^3) + (15 \times 16^4)$$
$$= 64 + 512 + 16{,}384 + 983{,}040$$
$$= 1{,}000{,}000$$

Nondenumerable Number Sets

A number set is nondenumerable if and only if there is *no way* that its elements can be arranged as a list, where each element is assigned a counting number defining its position in the list. Examples of nondenumerable number sets include the set of *irrational numbers*, the set of *real numbers*, the set of *imaginary numbers*, and the set of *complex numbers*. These types of numbers are used to express theoretical values in science and engineering.

IRRATIONAL NUMBERS

An irrational number cannot be expressed as the ratio of two integers. Examples of irrational numbers include:

- the length of the diagonal of a square that is 1 unit long on each edge (the square root of 2, roughly equal to 1.41421)
- the circumference-to-diameter ratio of a circle in a plane (commonly known as pi and symbolized π, roughly equal to 3.14159)

Irrational numbers are inexpressible in decimal-expansion form. When an attempt is made to express such a number in this form, the result is a decimal expression that is *nonterminating* and *nonrepeating*. No matter how many digits are specified to the right of the radix point, the expression is always an approximation, never the exact value.

The set of irrational numbers can be denoted S. This set is entirely disjoint from the set of rational numbers:

$$S \cap Q = \varnothing$$

This means that no rational number is irrational, and no irrational number is rational.

REAL NUMBERS

The set of real numbers, denoted R, is the union of the sets of rational and irrational numbers:

$$R = Q \cup S$$

For practical purposes, R can be denoted as the set of points on a continuous geometric line, as shown in Fig. 1-5. (In theoretical mathematics, the assertion that the points on a geometric line correspond one-to-one with the real numbers is known as the *Continuum Hypothesis*.) The real numbers are related to the rational numbers, the integers, and the natural numbers as follows:

$$N \subset Z \subset Q \subset R$$

The operations of addition, subtraction, multiplication, division, and exponentiation can be defined over the set of real numbers. If # represents any one of these operations and x and y are elements of R, then:

$$x \# y \in R$$

The only exception to this is that for division, y must not be equal to 0. Division by 0 is not defined within the set of real numbers.

TRANSFINITE CARDINAL NUMBERS

The cardinal numbers for infinite sets are denoted using the uppercase *aleph* ($\aleph$), the first letter in the Hebrew alphabet. The cardinality of the sets of natural

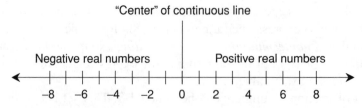

Fig. 1-5. The real numbers can be depicted as all the points on a continuous, solid, horizontal line. Displacement to the right corresponds to positive values, and displacement to the left corresponds to negative values.

numbers, integers, and rational numbers is called $\aleph_0$ (*aleph null*, *aleph nought*, or *aleph 0*). The cardinality of the sets of irrational and real numbers is called $\aleph_1$ (*aleph one* or *aleph 1*). These two quantities, $\aleph_0$ and $\aleph_1$, are known as *transfinite cardinal numbers*. They are expressions of "infinity."

Around the year 1900, the German mathematician Georg Cantor proved that $\aleph_0$ and $\aleph_1$ are not the same. This reflects the fact that the elements of the set of natural numbers can be paired off one-to-one with the elements of the sets of integers or rational numbers, but not with the elements of the sets of irrational numbers or real numbers. Any attempt to pair off the elements of *N* with the elements of *S*, or the elements of *N* and the elements of *R*, results in some elements of *S* or *R* being "left over" without corresponding elements in *N*. A simplistic, but interesting, way of saying this is that there are at least two "infinities," and they are not equal to each other!

IMAGINARY NUMBERS

The set of real numbers, and the operations defined above for the integers, give rise to some expressions that do not behave as real numbers. The best known example is the quantity j such that $j \times j = -1$. Thus, j is equal to the positive square root of -1. No real number has this property. This quantity j is known as the *unit imaginary number* or the *j operator*. Sometimes, in theoretical mathematics, j is denoted i.

The j operator can be multiplied by any real number x, called a *real-number coefficient*, and the result is an *imaginary number*. The coefficient x is written after j if x is positive or 0, and after $-j$ if x is negative. Examples are $j3$, $-j5$, and $-j2.787$. Numbers like this originally got the nickname "imaginary" because some people found it incredible that the square root of a negative real number could exist! But in pure mathematics, imaginary numbers are no more or less "imaginary" than real numbers.

The set *J* of all possible real-number multiples of j composes the entire set of imaginary numbers:

$$J = \{k \mid k = jx, \text{ where } x \in \mathbf{R}\}$$

For practical purposes, the set *J* can be depicted along a number line corresponding one-to-one with the real number line. By convention, the imaginary number line is oriented vertically, rather than horizontally (Fig. 1-6).

The sets of imaginary and real numbers have one element in common. That element is zero. When either j or $-j$ is multiplied by 0, the result is equal to the real number 0. Therefore, the intersection of the sets of imaginary and real numbers contains one element, namely 0. Formally we can write these statements like this:

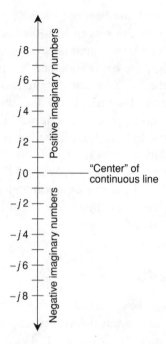

Fig. 1-6. The imaginary numbers can be depicted as all the points on a continuous, solid, vertical line. Upward displacement corresponds to positive imaginary values, and downward displacement corresponds to negative imaginary values.

$$j0 = -j0 = 0$$
$$\boldsymbol{J} \cap \boldsymbol{R} = \{0\}$$

COMPLEX NUMBERS

A *complex number* consists of the sum of two separate components, a real number and an imaginary number. The general form for a complex number c is:

$$c = a + jb$$

where a and b are real numbers. The set $\boldsymbol{C}$ of all complex numbers is thus defined as follows:

$$\boldsymbol{C} = \{c \mid c = a + jb, \text{ where } a \in \boldsymbol{R} \text{ and } b \in \boldsymbol{R}\}$$

If the real-number coefficient happens to be negative, then its *absolute value* (the value with the minus sign removed) is written following j, and a minus sign is used instead of a plus sign in the composite expression. So:

$$a + j(-b) = a - jb$$

Individual complex numbers can be depicted as points on a coordinate plane as shown in Fig. 1-7. The intersection point between the real and imaginary number lines corresponds to the value 0 on the real-number line and the value $j0$ on the imaginary-number line. (The real and imaginary zeroes are identical; that is, $0 = j0$. Therefore, they can both be represented by a single point.) Extrapolating the Continuum Hypothesis, the points on the so-called *complex-number plane* exist in a one-to-one correspondence with the elements of *C*.

The set of imaginary numbers is a proper subset of the set of complex numbers. The set of real numbers is also a proper subset of the set of complex numbers. Formally, we can write these statements as follows:

$$J \subset C$$
$$R \subset C$$

The sets of natural numbers (*N*), integers (*Z*), rational numbers (*Q*), real numbers (*R*), and complex numbers (*C*) can be related in a hierarchy of proper subsets:

$$N \subset Z \subset Q \subset R \subset C$$

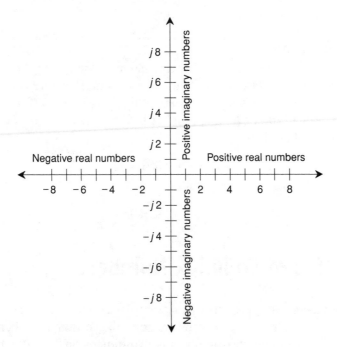

Fig. 1-7. The complex numbers can be depicted as points on a plane defined by the intersection of the real and the imaginary number lines at right angles.

PROBLEM 1-7

Earlier the statement was made that division by 0 was not defined over the set of real numbers. Yet, some texts treat expressions such as 1/0 as being equal to "infinity." This seems to make sense. The expression $1/x$, where x is a variable, gets larger and larger without limit as x approaches 0. Is 1/0 equal to "infinity?"

SOLUTION 1-7

The fact that the expression $1/x$ grows without limit as x approaches 0 does not logically imply that $1/x$ becomes "infinity" when x actually reaches 0. For us to be certain about that, we'd have to formally prove it. Even if we did that, we'd have to be sure what we meant by "infinity." Would we be talking about aleph 0 or aleph 1, or about some other sort of "infinity"?

PROBLEM 1-8

We have been told that $j \times j = -1$. This fact suggests that the square root of -1 is equal to j. What about the square root of some other negative real number, such as -4 or -100?

SOLUTION 1-8

The positive square root of any negative real number is equal to j times the positive square root of that real number. (There are negative square roots, too, but let's not worry about them right now.) If we let the positive square root of a real number be denoted as the 1/2 power of that real number, then:

$$(-4)^{1/2} = j \times 4^{1/2} = j2$$
$$(-100)^{1/2} = j \times 100^{1/2} = j10$$

Special Properties of Complex Numbers

Complex numbers have properties that are, in certain ways, similar to the properties of real numbers. But there are some big differences. Perhaps most significant, the set of complex numbers is two-dimensional (2D), while the set of real numbers is one-dimensional (1D). Complex numbers have two independent components, while real numbers consist of only one component.

EQUALITY OF COMPLEX NUMBERS

Let x_1 and x_2 be complex numbers such that:

$$x_1 = a_1 + jb_1$$
$$x_2 = a_2 + jb_2$$

These two complex numbers are *equal* if and only if their real components are equal and their imaginary components are equal:

$$x_1 = x_2 \text{ if and only if } a_1 = a_2 \text{ and } b_1 = b_2$$

OPERATIONS WITH COMPLEX NUMBERS

The operations of addition, subtraction, multiplication, division, and exponentiation are defined for the set of complex numbers as follows.

Complex addition: The real and imaginary parts are added independently. The general formula for the sum of two complex numbers is:

$$(a + jb) + (c + jd) = (a + c) + j(b + d)$$

Complex subtraction: The second complex number is multiplied by −1, and then the resulting two numbers are added. The general formula for the difference of two complex numbers is:

$$(a + jb) - (c + jd) = (a + jb) + [-1(c + jd)] = (a - c) + j(b - d)$$

Complex multiplication: The general formula for the product of two complex numbers is:

$$(a + jb)(c + jd) = ac + jad + jbc + j^2bd = (ac - bd) + j(ad + bc)$$

Complex division or ratio: The general formula for the quotient, or ratio, of two complex numbers is:

$$(a + jb) / (c + jd) = [(ac + bd) / (c^2 + d^2)] + j [(bc - ad) / (c^2 + d^2)]$$

The square brackets, while technically superfluous, are included to clarify the real and imaginary parts of the quotient. For the above formula to work, the denominator must not be equal to $0 + j0$. That means that c and d cannot both be equal to 0:

$$c + jd \neq 0 + j0$$

Complex exponentiation to a positive-integer power: This is symbolized by a superscript numeral. If $a + jb$ is an integer and c is a positive integer, then $(a + jb)^c$ is the result of multiplying $(a + jb)$ by itself c times.

COMPLEX CONJUGATES

Let x_1 and x_2 be complex numbers such that:

$$x_1 = a + jb$$
$$x_2 = a - jb$$

Then x_1 and x_2 are said to be *complex conjugates*, and they have the following two properties:

$$x_1 + x_2 = 2a$$
$$x_1 x_2 = a^2 + b^2$$

Conjugates are encountered when complex numbers are used in certain engineering applications. In particular, they appear in electronics when working with *impedance* in wireless communications systems. Impedance is a complex-number quantification of the opposition that a circuit offers to a high-frequency alternating current (AC).

MAGNITUDE AND DIRECTION

Complex numbers can be represented as *vectors*. A vector is a quantity having two independent properties: *magnitude* and *direction*. Any given complex number has a unique magnitude and a unique direction in a 2D coordinate system. The magnitude is the distance of the point $a + jb$ from the origin $0 + j0$. The direction is the angle that the vector subtends, expressed in a counterclockwise sense from the positive real-number axis. This principle is shown in Fig. 1-8.

The *absolute value* of a complex number $a + jb$, written $| a + jb |$, is the length of its vector in the complex plane, measured from the origin $(0,0)$ to the point (a,b). In the case of a pure real number $a + j0$:

$$| a + j0 | = a \text{ if } a \geq 0$$
$$| a + j0 | = -a \text{ if } a < 0$$

In the case of a pure imaginary number $0 + jb$:

$$| 0 + jb | = b \text{ if } b \geq 0$$
$$| 0 + jb | = -b \text{ if } b < 0$$

If a complex number is neither pure real nor pure imaginary, the absolute value is the length of the vector as shown in Fig. 1-9. This is derived from the *Theorem of Pythagoras* in basic plane geometry. The vector forms the *hypo-*

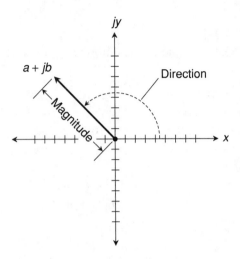

Fig. 1-8. A number in the complex plane can be defined as a vector, having a specific magnitude (or length) and a specific direction.

tenuse (longest side) of the right triangle. In the example of Fig. 1-9, this side is shown with length equal to c. The other two sides have lengths a and b. The formula for the absolute value of $a + jb$ in this case is:

$$| a + jb | = (a^2 + b^2)^{1/2}$$

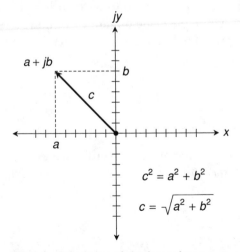

$$c^2 = a^2 + b^2$$

$$c = \sqrt{a^2 + b^2}$$

Fig. 1-9. Calculation of absolute value, or vector length, for a complex number. Here, the vector length is represented by c.

PROBLEM 1-9
Find the absolute value of the complex number $18 - j24$.

SOLUTION 1-9
Here, let $a = 18$ and $b = -24$. Using the above formula for the length of the vector, do the calculations as follows:

$$| a + jb | = (a^2 + b^2)^{1/2}$$
$$= [18^2 + (-24)^2]^{1/2}$$
$$= (324 + 576)^{1/2}$$
$$= 900^{1/2}$$
$$= 30$$

PROBLEM 1-10
How many complex numbers can exist with an absolute value of 6?

SOLUTION 1-10
There are infinitely many such complex numbers. The set of them all, shown as points in the complex-number plane, is as a circle of radius 6, centered at the origin $0 + j0$. There are infinitely many vectors corresponding to these complex numbers. All the vectors have lengths equal to 6, and point outward from the origin in all possible "compass directions" within the plane. Figure 1-10 shows a few such vectors, and the circle formed by the points corresponding to all complex numbers with absolute value equal to 6.

Quick Practice

Here are some practice problems that cover the material presented in this chapter. Solutions follow the problems.

PROBLEMS

1. Find $\{-3, -2, -1, 0, 1, 2, 3\} \cup \{0, 1, 2, 3, 4, 5, \ldots\}$.

2. Find $\{-3, -2, -1, 0, 1, 2, 3\} \cap \{0, 1, 2, 3, 4, 5, \ldots\}$.

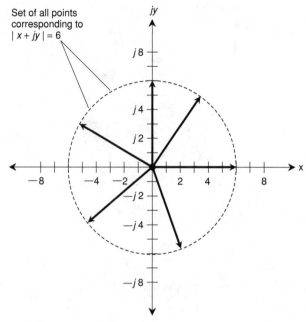

Fig. 1-10. Illustration for Problem 1-10.

3. Draw a Venn diagram showing two non-empty sets A and B, such that A is a proper subset of B.

4. Draw a Venn diagram showing two non-empty sets A and B, such that A and B are disjoint.

5. Convert the octal number 77 to decimal form.

SOLUTIONS

1. The union of two sets is the set of all elements belonging to either or both of the sets. Therefore:

$$\{-3, -2, -1, 0, 1, 2, 3\} \cup \{0, 1, 2, 3, 4, 5, \ldots\}$$
$$= \{-3, -2, -1, 0, 1, 2, 3, 4, 5, \ldots\}$$

2. The intersection of two sets is the set of all elements belonging to both sets. Therefore:

$$\{-3, -2, -1, 0, 1, 2, 3\} \cap \{0, 1, 2, 3, 4, 5, \ldots\} = \{0, 1, 2, 3\}$$

3. An example of two such sets is shown in Fig. 1-11.

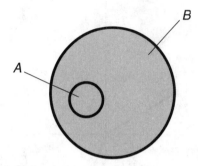

Fig. 1-11. Illustration for Quick
Practice Problem and Solution 3.

4. An example of two such sets is shown in Fig. 1-12.

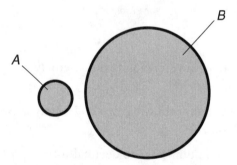

Fig. 1-12. Illustration for Quick Practice
Problem and Solution 4.

5. Working from right to left, octal 77 converts to decimal
    form as follows:

$$77 = (7 \times 8^0) + (7 \times 8^1)$$
$$= (7 \times 1) + (7 \times 8)$$
$$= 7 + 56$$
$$= 63$$

Quiz

This is an "open book" quiz. You may refer to the text in this chapter. A good score is 8 correct. Answers are in the back of the book.

1. What is the decimal equivalent of the largest possible 3-digit hexa-decimal number?
   (a) 2457
   (b) 4095
   (c) 8191
   (d) It is impossible to calculate this without more information.

2. Is $457/(-999)$ a rational number? If so, why? If not, why not?
   (a) Yes, because it is fully expressible in decimal form with a finite number of digits to the right of the radix point.
   (b) Yes, because it is the quotient of two integers.
   (c) No, because the denominator is a negative number.
   (d) No, because it is not expressible in decimal form at all, even with an infinite number of digits to the right of the radix point.

3. What is the value of $-j$ multiplied by itself?
   (a) 1
   (b) −1
   (c) j
   (d) $-j$

4. Let X and Y be sets. Suppose X is a proper subset of Y. Which of the following is impossible?
   (a) X has a finite number of elements, and Y has an infinite number of elements.
   (b) X and Y both have infinite numbers of elements.
   (c) X and Y both have finite numbers of elements.
   (d) X has an infinite number of elements, and Y has a finite number of elements.

5. Let X represent the set of all even integers (integers divisible by 2 without a remainder). Let Y represent the set of all integers. Which of the following statements is false?

(a) Sets X and Y are disjoint.
(b) $X \subseteq Y$
(c) $X \subset Y$
(d) $X \cap Y = X$

6. Working entirely in the octal number system, how would you write down the numeral representing the value of $1000 - 1$?

(a) 999
(b) 777
(c) 333
(d) 111

7. Suppose you have a binary number consisting of 3 digits. Then you place the digit 1 to the left of these, making a new, 4-digit binary number. The decimal equivalent of the new binary number is

(a) 8 times as large as the decimal equivalent of the original.
(b) 16 times as large as the decimal equivalent of the original.
(c) 8 larger than the decimal equivalent of the original.
(d) 16 larger than the decimal equivalent of the original.

8. Working entirely in the hexadecimal number system, how would you write down the numeral representing the value of $1000 - 3$?

(a) F97
(b) 99D
(c) F9D
(d) FFD

9. In the binary number system, what is $111 + 10$?

(a) 1001
(b) 1011
(c) 1100
(d) 1110

10. What is the product $(5 + j3) \times (-3 - j5)$, expressed in full as a complex number?

(a) $-15-j15$
(b) $15 + j15$
(c) $0 - j34$
(d) $0 + j0$

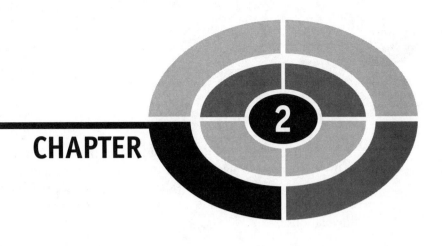

Principles of Calculation

Let's examine the properties of operations between and among numbers. You've likely seen most or all of this material before, but this chapter can serve as a refresher, especially if it has been a long time since you last worked with them.

Basic Principles

Several general principles, also called *laws*, are recognized for the operations of addition, subtraction, multiplication, and division for all real and complex numbers.

ADDITIVE IDENTITY

When 0 is added to any real number a, the sum is always equal to a. When $0 + j0$ is added to any complex number $a + jb$, the sum is always equal to $a + jb$. The numbers 0 and $0 + j0$ (identical in practice, but semantically distinct) are known as the real and complex *additive identity elements*. The following two equations hold:

$$a + 0 = a$$
$$(a + jb) + (0 + j0) = a + jb$$

MULTIPLICATIVE IDENTITY

When any real number a is multiplied by 1, the product is always equal to a. When any complex number $a + jb$ is multiplied by $1 + j0$, the product is always equal to $a + jb$. The numbers 1 and $1 + j0$ (identical in practice, but semantically distinct) are known as the real and complex *multiplicative identity elements*. The following two equations hold:

$$a \times 1 = a$$
$$(a + jb) \times (1 + j0) = a + jb$$

ADDITIVE INVERSES

For every real number a, there exists a unique real number $-a$ such that the sum of the two is equal to 0. For every complex number $a + jb$, there exists a unique complex number $-a - jb$ such that the sum of the two is equal to $0 + j0$. These pairs of numbers are known as the real and complex *additive inverses*. The following two equations hold:

$$a + (-a) = 0$$
$$(a + jb) + (-a - jb) = 0 + j0$$

MULTIPLICATIVE INVERSES

For every nonzero real number a, there exists a unique real number $1/a$ such that the product of the two is equal to 1. For every complex number $a + jb$ except

$0 + j0$, there exists a unique complex number $a/(a^2 + b^2) - jb/(a^2 + b^2)$ such that the product of the two is equal to $1 + j0$. These pairs of numbers are known as the real and complex *multiplicative inverses*. The following two equations hold:

$$a \times (1/a) = 1$$
$$(a + jb) \times [a/(a^2 + b^2) - jb/(a^2 + b^2)] = 1 + j0$$

COMMUTATIVE LAW OF ADDITION

When two real or complex numbers are added, it does not matter in which order the sum is performed. This is called the *commutative law of addition*. For all real numbers a and b, and for all complex numbers $a + jb$ and $c + jd$, the following two equations hold:

$$a + b = b + a$$
$$(a + jb) + (c + jd) = (c + jd) + (a + jb)$$

COMMUTATIVE LAW OF MULTIPLICATION

When two real or complex numbers are multiplied, it does not matter in which order the product is performed. This is called the *commutative law of multiplication*. For all real numbers a and b, and for all complex numbers $a + jb$ and $c + jd$, the following two equations hold:

$$ab = ba$$
$$(a + jb)(c + jd) = (c + jd)(a + jb)$$

ASSOCIATIVE LAW OF ADDITION

When three real or complex numbers are added, it does not matter how the addends are grouped. This is called the *associative law of addition*. For all real numbers a_1, a_2, and a_3, and for all complex numbers $a_1 + jb_1$, $a_2 + jb_2$, and $a_3 + jb_3$, the following two equations hold:

$$(a_1 + a_2) + a_3 = a_1 + (a_2 + a_3)$$
$$[(a_1 + jb_1) + (a_2 + jb_2)] + (a_3 + jb_3) = (a_1 + jb_1) + [(a_2 + jb_2) + (a_3 + jb_3)]$$

ASSOCIATIVE LAW OF MULTIPLICATION

When three real or complex numbers are multiplied, it does not matter how the factors are grouped. This is called the *associative law of multiplication*. For all real numbers a_1, a_2, and a_3, and for all complex numbers $a_1 + jb_1$, $a_2 + jb_2$, and $a_3 + jb_3$, the following two equations hold:

$$(a_1 a_2)a_3 = a_1(a_2 a_3)$$
$$[(a_1 + jb_1)(a_2 + jb_2)](a_3 + jb_3) = (a_1 + jb_1)[(a_2 + jb_2)(a_3 + jb_3)]$$

DISTRIBUTIVE LAW OF MULTIPLICATION OVER ADDITION

For all real numbers a_1, a_2, and a_3, and for all complex numbers $a_1 + jb_1$, $a_2 + jb_2$, and $a_3 + jb_3$, the following two equations hold. These equations express the *distributive law of multiplication over addition*:

$$a_1(a_2 + a_3) = a_1 a_2 + a_1 a_3$$
$$(a_1 + jb_1)[(a_2 + jb_2) + (a_3 + jb_3)] = (a_1 + jb_1)(a_2 + jb_2) + (a_1 + jb_1)(a_3 + jb_3)$$

PROBLEM 2-1

Demonstrate the validity of the commutative law of addition using three specific complex numbers.

SOLUTION 2-1

Consider $a = 2 + j3$ and $b = 4 - j5$. First, find the sum $a + b$:

$$a + b = (2 + j3) + (4 - j5)$$
$$= (2 + 4) + j(3 - 5)$$
$$= 6 + j(-2)$$
$$= 6 - j2$$

Then find the sum $b + a$:

$$b + a = (4 - j5) + (2 + j3)$$
$$= (4 + 2) + j(-5 + 3)$$
$$= 6 + j(-2)$$
$$= 6 - j2$$

These two sums are the same, which demonstrates the principle for these two particular numbers.

PROBLEM 2-2
What is the value of $4 \times (3 + 7)$? Compare this with the value of $(4 \times 3) + 7$.

SOLUTION 2-2
The value of the first expression is calculated using the distributive law of multiplication over addition:

$$4 \times (3 + 7) = (4 \times 3) + (4 \times 7)$$
$$= 12 + 28$$
$$= 40$$

The value of the second expression is calculated by first multiplying 4 by 3, and then adding 7 to the result:

$$(4 \times 3) + 7 = 12 + 7$$
$$= 19$$

Miscellaneous Principles

Here are some rules and definitions that apply to arithmetic operations for real and complex numbers, with the constraint that no denominator be equal to zero, and no denominator contain any variable that can attain a value that renders the denominator equal to zero.

ZERO NUMERATOR

If 0 is divided by any nonzero real number, the result is always equal to 0. If $0 + j0$ is divided by any complex number, the result is always equal to $0 + j0$. For all real numbers a such that $a \neq 0$, and for all complex numbers $a + jb$ such that $a + jb \neq 0 + j0$, the following two equations hold:

$$0/a = 0$$
$$0/(a + jb) = 0 + j0$$

ZERO DENOMINATOR

If any expression contains a denominator that is equal to 0, then that expression is undefined over the set of real and complex numbers. That is, for all real numbers a and all complex numbers $a + jb$, the following expressions are all undefined:

$$a/0$$
$$a/(0 + j0)$$
$$(a + jb)/0$$
$$(a + jb)/(0 + j0)$$

MULTIPLICATION BY ZERO

When any real number is multiplied by 0, the result is equal to 0. When any complex number is multiplied by $0 + j0$, the result is equal to $0 + j0$. For all real numbers a and all complex numbers $a + jb$, the following equations hold:

$$a \times 0 = 0$$
$$(a + jb) \times 0 = 0 + j0$$

ZEROTH POWER

The zeroth power of any nonzero real number is equal to 1. The zeroth power of any complex number is equal to $1 + j0$. For all real numbers a, where $a \neq 0$, and for all complex numbers $a + jb$, where $a + jb \neq 0 + j0$, the following two equations hold:

$$a^0 = 1$$
$$(a + jb)^0 = 1 + j0$$

The quantities 0^0 and $(0 + j0)^0$ are undefined.

POSITIVE INTEGER ROOTS

If x is a real or complex number and x is multiplied by itself n times to obtain another real or complex number y, then x is defined as the nth root of y. The following equations apply for all positive integers n:

$$x^n = y$$
$$x = y^{(1/n)}$$

FACTORIAL

If n is a positive integer, the value of $n!$ (n factorial) is the product of all positive integers less than or equal to n. The following equation applies for all positive integers n:

$$n! = 1 \times 2 \times 3 \times 4 \times ... \times n$$

In some texts, the value of $0!$ is defined as equal to 1 by default. The factorials of negative integers are not defined.

ARITHMETIC MEAN

Let $a_1, a_2, a_3, ...,$ and a_n be real numbers. The *arithmetic mean* (also known as the *average*) of $a_1, a_2, a_3, ...,$ and a_n is given by the following formula:

$$m_A = (a_1 + a_2 + a_3 + ... + a_n)/n$$

The arithmetic mean arises in such fields as meteorology, economics, statistics, and medicine. For example, the *mean temperature* can be defined for a particular period of time in a particular location, or the *mean age* of first-heart-attack victims in the United States can be used in a medical research paper.

GEOMETRIC MEAN

Let $a_1, a_2, a_3, ...,$ and a_n be real numbers. The *geometric mean* of $a_1, a_2, a_3, ...,$ and a_n is given by the following formula:

$$m_G = (a_1 a_2 a_3 ... a_n)^{1/n}$$

The geometric mean arises in the fields of physics and electronics. For example, the ideal *characteristic impedance* of a quarter-wave matching cable for a radio-frequency (RF) antenna system is equal to the geometric mean of the *input impedance* and the *load impedance*, assuming both of these impedances are *purely resistive*.

PRODUCT OF SIGNS

When numbers with plus (+) and minus (−) signs are multiplied, the following rules apply:

$$(+)(+) = (+)$$
$$(+)(-) = (-)$$
$$(-)(+) = (-)$$
$$(-)(-) = (+)$$

QUOTIENT OF SIGNS

When numbers with plus and minus signs are divided, the following rules apply:

$$(+)/(+) = (+)$$
$$(+)/(-) = (-)$$
$$(-)/(+) = (-)$$
$$(-)/(-) = (+)$$

POWER OF SIGNS

When numbers with signs are raised to a positive integer power n, the following rules apply:

$$(+)^n = (+)$$
$$(-)^n = (-) \text{ if } n \text{ is odd}$$
$$(-)^n = (+) \text{ if } n \text{ is even}$$

THE RECIPROCAL DEFINED

The *reciprocal* of a real number is equal to 1 divided by that number. The reciprocal of a complex number is equal to $1 + j0$ divided by that number.

RECIPROCAL OF RECIPROCAL

For all real numbers a such that $a \neq 0$, and for all complex numbers $a + jb$ such that $a + jb \neq 0 + j0$, the reciprocal of the reciprocal is always equal to the original number. The following two equations hold:

$$1/(1/a) = a$$
$$(1 + j0)/[(1 + j0)/(a + jb)] = a + jb$$

PROBLEM 2-3
What is the 1/5 power of 32?

SOLUTION 2-3
Using the formula above for positive integer roots, note that $32^{(1/5)}$ is the same thing as the 5th root of 32. This is equal to 2, because $2^5 = 2 \times 2 \times 2 \times 2 \times 2 = 32$.

PROBLEM 2-4
Find the arithmetic mean of 10 and 40. Compare the arithmetic mean with the geometric mean of these same two numbers.

SOLUTION 2-4
Using the above formula for arithmetic mean, we obtain:

$$m_A = (10 + 40)/2$$
$$= 50/2$$
$$= 25$$

The geometric mean is different:

$$m_G = (10 \times 40)^{1/2}$$
$$= 400^{1/2}$$
$$= 20$$

Advanced Principles

Here are some properties of arithmetic operations that you don't see as often as the previous ones. Nevertheless, they can be useful when manipulating equations.

PRODUCT OF SUMS

For all real or complex numbers w, x, y, and z, this formula can be used to find the product of $(w + x)$ and $(y + z)$:

$$(w + x)(y + z) = wy + wz + xy + xz$$

This formula can also be used to multiply factors containing differences. The difference between two quantities is considered to be equal to the first quantity plus the negative of the second quantity. Therefore:

$$(w - x)(y + z) = [w + (-x)] \, (y + z)$$
$$(w + x)(y - z) = (w + x) \, [y + (-z)]$$
$$(w - x)(y - z) = [w + (-x)] \, [y + (-z)]$$

DISTRIBUTIVE PROPERTY OF DIVISION OVER ADDITION

For all real numbers x, y, and z where $x \neq 0$, and for all complex numbers x, y, and z where $x \neq 0 + j0$, the following formula holds true:

$$(y + z)/x = y/x + z/x$$

CROSS-MULTIPLICATION

When two quotients are equal, their numerators and denominators can be cross-multiplied and the resulting products are equal. For all real numbers w, x, y, and z where $x \neq 0$ and $z \neq 0$, and for all complex numbers w, x, y, and z where $x \neq 0 + j0$ and $z \neq 0 + j0$, the following formulas hold true:

If $w/x = y/z$, then $wz = xy$

If $wz = xy$, then $w/x = y/z$

RECIPROCAL OF PRODUCT

The reciprocal of a product is equal to the product of the reciprocals. For all real numbers x and y where $x \neq 0$ and $y \neq 0$, and for all complex numbers x and y where $x \neq 0 + j0$ and $y \neq 0 + j0$, you can use these formulas:

$$1/(xy) = (1/x)(1/y)$$
$$(1 + j0)/(xy) = [(1 + j0)/x][(1 + j0)/y]$$

RECIPROCAL OF QUOTIENT

The reciprocal of a quotient is equal to the quotient expressed "upside-down." For all real numbers x and y where $x \neq 0$ and $y \neq 0$, and for all complex numbers x and y where $x \neq 0 + j0$ and $y \neq 0 + j0$, the following formulas apply:

$$1/(x/y) = y/x$$
$$(1 + j0)/(x/y) = y/x$$

PRODUCT OF QUOTIENTS

The product of two quotients is equal to the product of the numerators, divided by the product of the denominators. For all real numbers w, x, y, and z where $x \neq 0$ and $z \neq 0$, and for all complex numbers w, x, y, and z where $x \neq 0 + j0$ and $z \neq 0 + j0$, the following formula holds true:

$$(w/x)(y/z) = (wy)/(xz)$$

QUOTIENT OF PRODUCTS

For all real numbers w, x, y, and z where $y \neq 0$ and $z \neq 0$, and for all complex numbers w, x, y, and z where $y \neq 0 + j0$ and $z \neq 0 + j0$, either of these formulas can be used:

$$(wx)/(yz) = (w/y)(x/z)$$
$$(wx)/(yz) = (w/z)(x/y)$$

QUOTIENT OF QUOTIENTS

For all real numbers w, x, y, and z where $x \neq 0$, $y \neq 0$, and $z \neq 0$, and for all complex numbers w, x, y, and z where $x \neq 0 + j0$, $y \neq 0 + j0$, and $z \neq 0 + j0$, any of these formulas can be used:

$$(w/x)/(y/z) = (w/x)(z/y)$$
$$(w/x)/(y/z) = (w/y)(z/x)$$
$$(w/x)/(y/z) = (wz)/(xy)$$

SUM OF QUOTIENTS
(COMMON DENOMINATOR)

When two quotients have the same denominator, the sum of the quotients is equal to the sum of the numerators, divided by the common denominator. For all real numbers x, y, and z where $z \neq 0$, and for all complex numbers x, y, and z where $z \neq 0 + j0$, the following formula holds true:

$$x/z + y/z = (x + y)/z$$

SUM OF QUOTIENTS (GENERAL)

For all real numbers w, x, y, and z where $x \neq 0$ and $z \neq 0$, and for all complex numbers w, x, y, and z where $x \neq 0 + j0$ and $z \neq 0 + j0$, you can use this formula:

$$w/x + y/z = (wz + xy)/(xz)$$

RATIONAL-NUMBER EXPONENTS

Let x be a real or complex number. Let y be a rational number such that $y = a/b$, where a and b are integers and $b \neq 0$. The following formula can be used to find x to the yth power:

$$x^y = x^{a/b} = (x^a)^{(1/b)} = [x^{(1/b)}]^a$$

NEGATIVE EXPONENTS

Let x be a nonzero real or complex number. Let y be a rational number. The following formula can be used to find x to the $-y$th power:

$$x^{(-y)} = (1/x)^y = 1/x^y$$

SUM OF EXPONENTS

Let x be a complex number. Let y and z be rational numbers. The following formula can be used to find x to the $(y + z)$th power:

$$x^{(y+z)} = x^y x^z$$

DIFFERENCE OF EXPONENTS

Let x be a real or complex number, with the constraint that $x \neq 0$. Let y and z be rational numbers. The following formula can be used to find x to the $(y - z)$th power:

$$x^{(y-z)} = x^y / x^z$$

PRODUCT OF EXPONENTS

Let x be a real or complex number. Let y and z be rational numbers. The following formula can be used to find x to the (yz)th power:

$$x^{yz} = (x^y)^z = (x^z)^y$$

QUOTIENT OF EXPONENTS

Let x be a real or complex number. Let y and z be rational numbers, where $z \neq 0$. The following formula can be used to find x to the (y/z)th power:

$$x^{y/z} = (x^y)^{(1/z)} = [x^{(1/z)}]^y$$

POWERS OF SUM

Let x and y be real or complex numbers. The following formulas can be used to find the square, cube, and fourth power of the sum $(x + y)$:

$$(x + y)^2 = x^2 + 2xy + y^2$$
$$(x + y)^3 = x^3 + 3x^2y + 3xy^2 + y^3$$
$$(x + y)^4 = x^4 + 4x^3y + 6x^2y^2 + 4xy^3 + y^4$$

POWERS OF DIFFERENCE

Let x and y be real or complex numbers. Then the following formulas can be used to find the square, cube, and fourth power of the difference $(x - y)$:

$$(x - y)^2 = x^2 - 2xy + y^2$$
$$(x - y)^3 = x^3 - 3x^2y + 3xy^2 - y^3$$
$$(x - y)^4 = x^4 - 4x^3y + 6x^2y^2 - 4xy^3 + y^4$$

PROBLEM 2-5

What is the value of $4^{(-3)}$? Compare this with the value of $4^{(-1/3)}$. Express the answers in decimal form to several figures.

SOLUTION 2-5

Using the above formula for negative powers to find the −3rd power of 4:

$$4^{(-3)} = 1/(4^3) = 1/(4 \times 4 \times 4)$$
$$= 1/64$$
$$= 0.015625$$

The quantity $4^{(-1/3)}$ is calculated using the formula for negative powers first, and then the formula for positive integer roots:

$$4^{(-1/3)} = 1/[4^{(1/3)}]$$
$$= 1/(1.5874)$$
$$= 0.62996$$

PROBLEM 2-6

Find the value of $(2 + n)^2$, and compare it with the value of $(2 - n)^2$, where n is a variable.

SOLUTION 2-6

Use the previous formula for the power of a sum to find the first value. Substitute 2 for x and n for y:

$$(2 + n)^2 = 2^2 + (2 \times 2 \times n) + n^2$$
$$= 4 + 4n + n^2$$

Use the previous formula for the power of a difference to find the second value. Again, substitute 2 for x and n for y:

$$(2 - n)^2 = 2^2 - (2 \times 2 \times n) + n^2$$
$$= 4 - 4n + n^2$$

Approximation and Precedence

Numbers in the real world are not always exact. This is especially true in observational science and in engineering. Often, we must approximate. There are two ways of doing this: *truncation* (straightforward but less accurate) and *rounding* (a little trickier, but more accurate).

TRUNCATION

The process of truncation involves the deletion of all the numerals to the right of a certain point in the decimal part of an expression. Some electronic calculators use truncation to fit numbers within their displays. For example, the number 3.830175692803 can be shortened as follows, depending on the number of digits desired in the outcome:

$$3.830175692803$$
$$\approx 3.83017569280$$
$$\approx 3.8301756928$$
$$\approx 3.830175692$$
$$\approx 3.83017569$$
$$\approx 3.8301756$$
$$\approx 3.830175$$
$$\approx 3.83017$$
$$\approx 3.8301$$
$$\approx 3.830$$
$$\approx 3.83$$
$$\approx 3.8$$
$$\approx 3$$

The wavy equality symbol ($\approx$) means "is approximately equal to."

ROUNDING

Rounding is the preferred method of rendering numbers in shortened form. In this process, when a given digit (call it r) is deleted at the right-hand extreme of an expression, the digit q to its left (which becomes the new r after the old r is deleted) is not changed if $0 \leq r \leq 4$. If $5 \leq r \leq 9$, then q increases by 1 (round it up). Most electronic calculators use rounding. If rounding is used, the number 3.830175692803 can be shortened as follows, depending on the number of digits desired in the outcome:

$$3.830175692803$$
$$\approx 3.83017569280$$
$$\approx 3.8301756928$$
$$\approx 3.830175693$$
$$\approx 3.83017569$$

≈ 3.8301757

≈ 3.830176

≈ 3.83018

≈ 3.8302

≈ 3.830

≈ 3.83

≈ 3.8

≈ 4

PRECEDENCE

Mathematicians agree on a certain order in which operations should be performed when they appear together in an expression. This prevents confusion and ambiguity. When diverse operations appear in an expression, and if you need to simplify that expression, perform the operations in the following sequence:

- Simplify all expressions within parentheses, brackets, and braces from the inside out.
- Perform all exponential operations.
- Perform all products.
- Perform all quotients.
- For all quantities x and y, consider a difference $x - y$ as a sum $x + (-y)$.
- Perform all sums, proceeding from left to right.

Here are two examples of the above rules of precedence. Note that the order of the numerals and operations is the same in each case, but the groupings differ.

$$[(2 + 3)(-3 - 1)^2]^2 = [5 \times (-4)^2]^2$$
$$= (5 \times 16)^2$$
$$= 80^2$$
$$= 6400$$

$$[(2 + 3 \times (-3) - 1)^2]^2 = [(2 + (-9) - 1)^2]^2$$
$$= (-8^2)^2$$
$$= 64^2$$
$$= 4096$$

Suppose you're given a complicated expression and there are no parentheses, brackets, or braces in it. This is not ambiguous if the above mentioned rules are followed. Consider this example:

$$z = -3x^3 + 4x^2y - 12xy^2 - 5y^3$$

If this is written with parentheses, brackets, and braces to emphasize the rules of precedence, it looks like this:

$$z = [-3(x^3)] + \{4[(x^2)y]\} - \{12[x(y^2)]\} - [5(y^3)]$$

Because we have agreed on the rules of precedence, we can do without the parentheses, brackets, and braces. Nevertheless, if there is any doubt about a crucial equation, you should use a couple of unnecessary parentheses rather than risk making a calculation error.

PROBLEM 2-7
Truncate the value of the constant pi (π), which represents the ratio of the circumference of a circle to its diameter in plane geometry, in steps from 10 digits down to six digits. Then round it off in steps from 10 digits down to six digits.

SOLUTION 2-7
First, find a reference that shows π to at least 10 digits. Most scientific calculators, including the program in the computer operating system Windows XP, have a "pi" key. This key gives the following sequence for the first 10 digits of π:

$$\pi = 3.141592653$$

Truncating in steps down to six digits, we get this sequence of values:

$$3.141592653$$
$$\approx 3.14159265$$
$$\approx 3.1415926$$
$$\approx 3.141592$$
$$\approx 3.14159$$

Rounding in steps down to six digits produces the same end result, although a couple of the intermediate numbers are different:

$$3.141592653$$
$$\approx 3.14159265$$
$$\approx 3.1415927$$
$$\approx 3.141593$$
$$\approx 3.14159$$

PROBLEM 2-8

What is the value of $2 + 3 \times 4 + 5$?

SOLUTION 2-8

First, perform the multiplication operation, obtaining the expression $2 + 12 + 5$. Then add the numbers, obtaining the final value 19. Therefore:

$$2 + 3 \times 4 + 5 = 19$$

Quick Practice

Here are some practice problems that cover the material presented in this chapter. Solutions follow the problems.

PROBLEMS

1. Find the arithmetic mean of 3, 4, and 20.

2. Find the geometric mean of 0, 6, and 71.

3. Find the value of 8 factorial.

4. Find the value of 6 to the 0th power (6^0).

5. Round off $e = 2.718281828459\ldots$ in steps down to four digits.

SOLUTIONS

1. To find the arithmetic mean, calculate as follows:

$$(3 + 4 + 20)/3 = 27/3$$
$$= 9$$

2. To find the geometric mean, calculate as follows:

$$(0 \times 6 \times 71)^{(1/3)} = 0^{(1/3)}$$
$$= 0$$

3. To find 8 factorial, calculate as follows:

$$8! = 1 \times 2 \times 3 \times 4 \times 5 \times 6 \times 7 \times 8$$
$$= 40,320$$

4. The 0th power of any nonzero number is equal to 1. Therefore, $6^0 = 1$.

5. Here is the value of e as it is repeatedly rounded off:

$$2.718281828459$$
$$\approx 2.71828182846$$
$$\approx 2.7182818285$$
$$\approx 2.718281829$$
$$\approx 2.71828183$$
$$\approx 2.7182818$$
$$\approx 2.718282$$
$$\approx 2.71828$$
$$\approx 2.7183$$
$$\approx 2.718$$

Quiz

This is an "open book" quiz. You may refer to the text in this chapter. A good score is 8 correct. Answers are in the back of the book.

1. Using the product-of-sums rule, what is another expression for $(x + 2)(y - 2)$?
    (a) $xy + 2x + 2y + 4$
    (b) $xy - 2x + 2y + 4$
    (c) $xy - 2x + 2y - 4$
    (d) $xy - 2x - 2y - 4$

2. Using the product-of-sums rule "in reverse," what is another expression of the equation $x^2 + 8x + 16$?
    (a) $(x + 4)(x - 4)$
    (b) $(x^2 + 4)(x^2 - 4)$
    (c) $(x + 4)^2$
    (d) $(x - 4)^2$

3. The geometric and arithmetic means of x and y are the same if and only if

    (a) $(x + y)/2 = (xy)^{1/2}$

    (b) $(x + y)^2 = 1$

    (c) $x^2 + 2xy + y^2 = 1$

    (d) Forget it! This can never happen.

4. The product $(j + 1)(j - 1)$ is equal to

    (a) 2

    (b) 1

    (c) −1

    (d) −2

5. The product $(1 + j)(1 - j)$ is equal to

    (a) 2

    (b) 1

    (c) −1

    (d) −2

6. When you want to find the arithmetic or geometric mean of two numbers, it doesn't matter which number is expressed "first" and which number is expressed "second." This arises from the fact that addition and multiplication are both

    (a) associative.

    (b) commutative.

    (c) distributive.

    (d) complex.

7. Which of the following expressions is not defined for any real or complex value of x?

    (a) $(3x + 3)/2$

    (b) $x^2 + 10x + 100$

    (c) $x^2/(x - 3)$

    (d) $x^2/[3(x - x)]$

8. In which of the following expressions must we place a constraint on the value of x, in order to make sure that the expression is defined?

    (a) In the expression $(3x + 3)/2$. Here, the value of x must not be equal to 0.

    (b) In the expression $x^2 + 10x + 100$. Here, the value of x must not be equal to −10.

(c) In the expression $x^2/(x-3)$. Here, the value of x must not be equal to 3.

    (d) In the expression $x^2/[3(x-x)]$. Here, the value of x must be negative.

9. The expression $p^a p^b$ can be rewritten as

    (a) $p^{(ab)}$

    (b) $p^{(a+b)}$

    (c) $p^{(a-b)}$

    (d) $p^{(a/b)}$

10. Suppose you are confronted with the following expression that does not contain any parentheses or brackets to tell you the order in which operations should be done:

$$26 \times 6 \times 5^2 + 7 \times 8 \times 2 - 3/4 - 8/7$$

Which operation should you do first, in order to follow the rules of precedence?

    (a) Multiply 26 by 6.

    (b) Divide 8 by 7.

    (c) Square 5.

    (d) It doesn't make any difference what you do first.

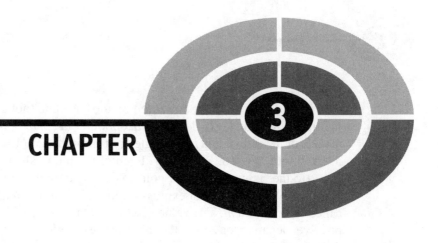

Scientific Notation

In engineering and the physical sciences, huge or tiny quantities can be unwieldy when written out as ordinary decimal numerals. *Scientific notation* provides a "shortcut" method of expressing such quantities, usually as approximations.

Powers of 10

The most common way to denote extreme quantities is a scheme in which they are portrayed as real-number multiples of integer powers of 10.

STANDARD FORM

A numeral in *standard scientific notation* (also called the American form) is written as follows:

$$m.n_1n_2n_3\ldots n_p \times 10^z$$

where the dot (.) is a period, written on the base line (not a raised dot indicating multiplication), and is called the *radix point* or *decimal point*. The numeral m (to the left of the radix point) is a single digit from the set {1, −1, 2, −2, 3, −3, 4, −4, 5, −5, 6, −6, 7, −7, 8, −8, 9, −9}. Each of the numerals n_1, n_2, n_3, and so on up to n_p (to the right of the radix point) is a single digit from the set {0, 1, 2, 3, 4, 5, 6, 7, 8, 9}. The decimal expression to the left of the multiplication symbol is called the *coefficient*. The value z, which is the power of 10, can be any integer: positive, negative, or zero. Here are some examples of numbers written in standard scientific notation:

$$7.63 \times 10^8$$
$$-4.10015 \times 10^{-15}$$
$$4.000 \times 10^0$$

ALTERNATIVE FORM

In certain countries, and in some scientific and technical papers and books, a variation on the above theme is used: *alternative scientific notation* (sometimes called the European form). This system requires that $m = 0$. When the above quantities are expressed this way, they appear as decimal fractions larger than 0 but less than 1, and the value of the exponent is increased by 1 compared with the value of the exponent for the same number in standard scientific notation. In alternative scientific notation, the above three quantities would be expressed like this:

$$0.763 \times 10^9$$
$$-0.410015 \times 10^{-14}$$
$$0.4000 \times 10^1$$

Note that when a negative exponent is "increased by 1," it becomes "1 less negatively." For example, when −15 is increased by 1, it becomes −14, not −16.

THE "TIMES SIGN"

The multiplication sign in scientific notation can be denoted in various ways. Most scientists in America use the cross symbol (×), as in the examples shown above. But a small dot raised above the base line (·) is sometimes used to represent multiplication in scientific notation. When written that way, the above numbers look like this in the standard power-of-10 form:

$$7.63 \cdot 10^8$$
$$-4.10015 \cdot 10^{-15}$$
$$4.000 \cdot 10^0$$

This small dot should not be confused with a radix point.

A small dot symbol is preferred when multiplication is required to express the dimensions of a physical unit. An example is the kilogram-meter per second squared, which is symbolized $kg \cdot m/s^2$ or $kg \cdot m \cdot s^{-2}$.

Another alternative multiplication symbol in scientific notation is the asterisk (*). You will occasionally see numbers written like this in standard scientific notation:

$$7.63 * 10^8$$
$$-4.10015 * 10^{-15}$$
$$4.000 * 10^0$$

PLAIN-TEXT EXPONENTS

Once in awhile, you will have to express numbers in scientific notation using plain, unformatted text. This is the case when transmitting information within the body of an e-mail message (rather than as an attachment). Some calculators and computers use this system in their displays. An uppercase letter E indicates that the quantity immediately before it is to be multiplied by a power of 10, and that power is written immediately after the E. In this format, the above quantities are written:

$$7.63E8$$
$$-4.10015E-15$$
$$4.000E0$$

In an alternative format, the exponent is always written with two numerals, and always includes a plus sign or a or minus sign, so the above expressions appear as:

$$7.63E+08$$
$$-4.10015E-15$$
$$4.000E+00$$

Still another alternative is the use of an asterisk to indicate multiplication, and the symbol ^ to indicate a superscript, so the expressions look like this:

$$7.63 * 10^8$$
$$-4.10015 * 10^{-15}$$
$$4.000 * 10^0$$

In all of these examples, the numerical values represented are identical. Respectively, if written out in full, they are:

$$763,000,000$$
$$-0.00000000000000410015$$
$$4.000$$

ORDERS OF MAGNITUDE

Consider the following two extreme numbers:

$$2.55 \times 10^{45,589}$$
$$-9.8988 \times 10^{-7,654,321}$$

Imagine the task of writing either of these numbers out in ordinary decimal form! In the first case, you would have to write the numerals 2, 5, and 5 in that order, and then follow them with a string of 45,587 zeroes. In the second case, you'd have to write a minus sign, then the numeral 0, then a radix point, then a string of 7,654,320 zeroes, and finally the numerals 9, 8, 9, 8, and 8, in that order.

Now consider these two numbers:

$$2.55 \times 10^{45,592}$$
$$-9.8988 \times 10^{-7,654,318}$$

Both of these numbers are 1000 times larger than the original two, because both exponents are larger by 3. The exponent 45,592 is 3 larger than the exponent 45,589, and the exponent −7,654,318 is 3 larger than the exponent −7,654,321. (Again, remember that numbers grow larger in the mathematical sense as they become more positive or less negative.) The second pair of numbers is therefore 3 *orders of magnitude* larger than the first pair of numbers.

The order-of-magnitude concept makes it possible to construct number lines, charts, and graphs with scales that cover large ranges or spans. Three examples are shown in Fig. 3-1. Drawing A shows a *logarithmic-scale* number line spanning 3 orders of magnitude, from 10^0 (1) to 10^3 (1000). Illustration B shows a logarithmic-scale number line spanning 10 orders of magnitude, from 10^{-3} (0.001) to 10^7 (10,000,000). Illustration C shows a coordinate system with a

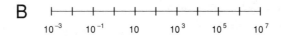

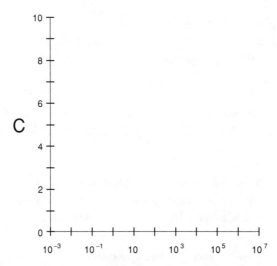

Fig. 3-1. At A, a number line spanning 3 orders of magnitude.
At B, a number line spanning 10 orders of magnitude. At C, a
coordinate system whose horizontal scale spans 10 orders of magnitude, and whose vertical scale extends linearly from 0 to 10.

logarithmic-scale horizontal axis spanning 10 orders of magnitude (from 10^{-3} to
10^{7}), and with a *linear-scale* vertical axis spanning values from 0 to 10.

WHEN TO USE SCIENTIFIC NOTATION

In formal technical documents, scientific notation is used only when the power
of 10 is large or small. If the exponent is between −2 and 2 inclusive, numbers
are written out in plain decimal form as a rule. If the exponent is −3 or 3, numbers are sometimes written out in plain decimal form, and are sometimes written in scientific notation. If the exponent is −4 or smaller, or if it is 4 or larger,
values are expressed in scientific notation as a rule. (In number lines and graphs,
exceptions are sometimes made for consistency, as is the case in Fig. 3-1.)

Some calculators, when set for scientific notation, display all numbers that way, even when it is not strictly necessary. This can be confusing, especially when the power of 10 is zero and the calculator is set to display many digits. Most people understand the expression 8.407 more easily than 8.407000000E+00, for example, even though they represent the same number.

PREFIX MULTIPLIERS

Special verbal prefixes, known as *prefix multipliers*, are used in the physical sciences and in engineering to express orders of magnitude. Table 3-1 shows the prefix multipliers, and their symbols, for factors ranging form 10^{-24} to 10^{24}.

PROBLEM 3-1
By how many orders of magnitude does a *gigahertz* differ from a *kilohertz*? (The *hertz* is a unit of frequency, equivalent to a cycle per second.)

SOLUTION 3-1
Refer to Table 3-1. A gigahertz represents 10^9 hertz, and a kilohertz represents 10^3 hertz. The exponents differ by 6. Therefore, a gigahertz differs from a kilohertz by 6 orders of magnitude.

Table 3-1. Power-of-10 prefix multipliers and their symbols.

Designator	Symbol	Multiplier	Designator	Symbol	Multiplier
yocto-	y	10^{-24}	deka-	da or D	10^1
zepto-	z	10^{-21}	hecto-	h	10^2
atto-	a	10^{-18}	kilo-	K or k	10^3
femto-	f	10^{-15}	mega-	M	10^6
pico-	p	10^{-12}	giga-	G	10^9
nano-	n	10^{-9}	tera-	T	10^{12}
micro-	μ or mm	10^{-6}	peta-	P	10^{15}
milli-	m	10^{-3}	exa-	E	10^{18}
centi-	c	10^{-2}	zetta-	Z	10^{21}
deci-	d	10^{-1}	yotta-	Y	10^{24}
(none)	—	10^0			

PROBLEM 3-2
What, if anything, is wrong with the number 971.82×10^5 as an expression in standard scientific notation?

SOLUTION 3-2
This is a legitimate number, but it is not written in the correct format for scientific notation. The number to the left of the multiplication symbol should be at least 1, but smaller than 10. To convert the number to the proper format, first divide the portion to the left of the multiplication symbol by 100, so it becomes 9.7182. Then multiply the portion to the right of the multiplication symbol by 100, increasing the exponent by 2 so it becomes 10^7. This produces the same numerical value but in the correct format for standard scientific notation: 9.7182×10^7.

Calculations in Scientific Notation

Let's see how scientific notation works when performing simple calculations involving common arithmetic operations.

MULTIPLICATION

When numbers are multiplied in scientific notation, the coefficients are multiplied by each other. Then the exponents are added. Finally, the product is reduced to standard form. Here are three examples:

$$(3.045 \times 10^5) \times (6.853 \times 10^6) = (3.045 \times 6.853) \times (10^5 \times 10^6)$$
$$= 20.867385 \times 10^{5+6}$$
$$= 20.867385 \times 10^{11}$$
$$= 2.0867385 \times 10^{12}$$

$$(3.045 \times 10^{-4}) \times (-6.853 \times 10^{-7}) = [3.045 \times (-6.853)] \times (10^{-4} \times 10^{-7})$$
$$= -20.867385 \times 10^{-4+(-7)}$$
$$= -20.867385 \times 10^{-11}$$
$$= -2.0867385 \times 10^{-10}$$

$$(-3.045 \times 10^5) \times (-6.853 \times 10^{-7}) = [(-3.045) \times (-6.853)] \times (10^5 \times 10^{-7})$$
$$= 20.867385 \times 10^{5-7}$$
$$= 20.867385 \times 10^{-2}$$
$$= 2.0867385 \times 10^{-1}$$
$$= 0.20867385$$

This last number is written out in plain decimal form because the exponent is between −2 and 2 inclusive.

DIVISION

When numbers are divided in scientific notation, the coefficients are divided by each other. Then the exponents are subtracted. Finally, the quotient is reduced to standard form. Here are three examples of how division is done in scientific notation:

$$(3.045 \times 10^5)/(6.853 \times 10^6) = (3.045/6.853) \times (10^5/10^6)$$
$$\approx 0.444331 \times 10^{5-6}$$
$$= 0.444331 \times 10^{-1}$$
$$= 0.0444331$$

$$(3.045 \times 10^{-4})/(-6.853 \times 10^{-7}) = [3.045/(-6.853)] \times (10^{-4}/10^{-7})$$
$$\approx -0.444331 \times 10^{-4-(-7)}$$
$$= -0.444331 \times 10^3$$
$$= -4.44331 \times 10^2$$
$$= -444.331$$

$$(-3.045 \times 10^5)/(-6.853 \times 10^{-7}) = [(-3.045)/(-6.853)] \times (10^5/10^{-7})$$
$$\approx 0.444331 \times 10^{5-(-7)}$$
$$= 0.444331 \times 10^{12}$$
$$= 4.44331 \times 10^{11}$$

The numbers here do not divide out neatly, so the decimal-format portions are approximated. The "wavy" or "squiggly" equals signs in the second lines of the preceding three calculation sequences mean "is approximately equal to."

EXPONENTIATION

When a number is raised to a power in scientific notation, both the coefficient and the power of 10 itself must be raised to that power, and the result multiplied. Consider this example:

$$(4.33 \times 10^5)^3 = (4.33)^3 \times (10^5)^3 = 81.182737 \times 10^{5 \times 3}$$
$$= 81.182737 \times 10^{15}$$
$$= 8.1182727 \times 10^{16}$$

Consider another example, in which the power of 10 is negative:

$$(5.27 \times 10^{-4})^2 = (5.27)^2 \times (10^{-4})^2 = 27.7729 \times 10^{-4 \times 2}$$
$$= 27.7729 \times 10^{-8}$$
$$= 2.77729 \times 10^{-7}$$

TAKING ROOTS

To find the root of a number in scientific notation, think of the root as a fractional exponent. The square root is equivalent to the 1/2 power. The cube root is the same thing as the 1/3 power. In general, the nth root of a number (where n is a positive integer) is the same thing as the 1/n power. When roots are regarded this way, it is easy to multiply things out in exactly the same way as is done with whole-number exponents. Here is an example:

$$(5.27 \times 10^{-4})^{1/2} = (5.27)^{1/2} \times (10^{-4})^{1/2} \approx 2.2956 \times 10^{-4 \times (1/2)}$$
$$\approx 2.2956 \times 10^{-2}$$
$$= 0.02956$$

ADDITION

Scientific notation is awkward when adding up sums, unless all of the addends are expressed to the same power of 10. Here are three examples:

$$(3.045 \times 10^5) + (6.853 \times 10^6) = 304{,}500 + 6{,}853{,}000 = 7{,}157{,}500$$
$$= 7.1575 \times 10^6$$

$$(3.045 \times 10^{-4}) + (6.853 \times 10^{-7}) = 0.0003045 + 0.0000006853$$
$$= 0.0003051853$$
$$= 3.051853 \times 10^{-4}$$

$$(3.045 \times 10^5) + (6.853 \times 10^{-7}) = 304{,}500 + 0.0000006853$$
$$= 304{,}500.0000006853$$
$$= 3.045000000006853 \times 10^5$$

SUBTRACTION

Subtraction follows the same basic rules as addition. It helps to convert the numbers to ordinary decimal format before subtracting:

$$(3.045 \times 10^5) - (6.853 \times 10^6) = 304{,}500 - 6{,}853{,}000$$
$$= -6{,}548{,}500$$
$$= -6.548500 \times 10^6$$

$$(3.045 \times 10^{-4}) - (6.853 \times 10^{-7}) = 0.0003045 - 0.0000006853$$
$$= 0.0003038147$$
$$= 3.038147 \times 10^{-4}$$

$$(3.045 \times 10^5) - (6.853 \times 10^{-7}) = 304{,}500 - 0.0000006853$$
$$= 304{,}499.9999993147$$
$$= 3.044999999993147 \times 10^5$$

PROBLEM 3-3
State the generalized rule for multiplication in scientific notation, using the variables u and v to represent the coefficients and the variables m and n to represent the exponents.

SOLUTION 3-3
Let u and v be real numbers greater than or equal to 1 but less than 10, and let m and n be integers. Then:

$$(u \times 10^m)(v \times 10^n) = uv \times 10^{m+n}$$

PROBLEM 3-4
State the generalized rule for division in scientific notation, using the variables u and v to represent the coefficients and the variables m and n to represent the exponents.

SOLUTION 3-4
Let u and v be real numbers greater than or equal to 1 but less than 10, and let m and n be integers. Then:

$$(u \times 10^m)/(v \times 10^n) = u/v \times 10^{m-n}$$

PROBLEM 3-5
State the generalized rule for exponentiation in scientific notation, using the variable u to represent the coefficient, and the variables m and n to represent the exponents.

SOLUTION 3-5
Let u be a real number greater than or equal to 1 but less than 10, and let m and n be integers. Then:

$$(u \times 10^m)^n = u^n \times 10^{mn}$$

Significant Figures

The number of *significant figures*, also called *significant digits*, in an expression indicates the degree of accuracy to which we know a numerical value, or to which we have measured, or can measure, a quantity.

MULTIPLICATION, DIVISION, AND EXPONENTIATION

When multiplication, division, or exponentiation is done using scientific notation, the number of significant figures in the final calculation result cannot legitimately be greater than the number of significant figures in the least exact expression.

Consider the two numbers $x = 2.453 \times 10^4$ and $y = 7.2 \times 10^7$. The following is a perfectly valid statement if the numerical values are exact:

$$xy = 2.453 \times 10^4 \times 7.2 \times 10^7 = 2.453 \times 7.2 \times 10^{11}$$
$$= 17.6616 \times 10^{11}$$
$$= 1.76616 \times 10^{12}$$

But if x and y represent measured quantities, as is nearly always the case in experimental science and engineering, the above statement needs qualification. We must pay close attention to how much accuracy we claim.

HOW ACCURATE ARE WE?

When you see a product or quotient containing quantities expressed in scientific notation, count the number of single digits in the coefficients of each number. Then take the smallest number of digits. This is the number of significant figures you can claim in the final answer or solution.

In the previous example, there are four single digits in the coefficient of x, and two single digits in the coefficient of y. So you must round off the answer, which appears to contain six significant figures, to two significant figures. It is important to use rounding, and not truncation, as follows:

$$xy = 2.453 \times 10^4 \times 7.2 \times 10^7$$
$$= 1.8 \times 10^{12}$$

In situations of this sort, if you insist on being rigorous, you can use approximate-equality symbols (the wavy ones) throughout, because you are always dealing with approximate values. But most folks are content to use ordinary equality symbols. It is universally understood that physical measurements are inherently inexact.

Suppose you want to find the quotient x/y instead of the product xy? Proceed as follows:

$$x/y = (2.453 \times 10^4)/(7.2 \times 10^7)$$
$$= (2.453/7.2) \times 10^{-3}$$
$$= 0.3406944444\ldots \times 10^{-3}$$
$$= 3.406944444\ldots \times 10^{-4}$$
$$= 3.4 \times 10^{-4}$$

WHAT ABOUT 0?

Sometimes, when you make a calculation, you will get an answer that lands on a neat, seemingly whole-number value. Consider $x = 1.41421$ and $y = 1.41422$. Both of these have six significant figures. The product, taking significant figures into account, is:

$$xy = 1.41421 \times 1.41422$$
$$= 2.0000040662$$
$$= 2.00000$$

This appears to be exactly equal to 2. But in the real world (the measurement of a physical quantity, for example), the presence of five zeros after the radix point indicates an uncertainty of up to plus-or-minus 0.000005 (written ±0.000005). When we claim a certain number of significant figures, 0 is as important as any other digit.

WHAT ABOUT EXACT VALUES?

In some cases, values in physical formulas are given as exact. An example of this is the equation for the area of a triangle, A, in terms of its base length b and its height h:

$$A = bh/2$$

In this formula, 2 is a constant, and its value is exact. It can therefore have as many significant figures as we want, depending on the number of significant figures we are given in the initial values. We can call it 2.0000000…, in effect claiming an infinite number of significant figures, all of which, except for the initial digit, are 0.

Sometimes there are constants in equations whose values can be taken as exact, but which we must nevertheless round off when we want to assign it a certain number of significant figures. A common example of this is π (pi), which is the ratio of the circumference of a circle to its straight-line diameter. This has a theoretically exact value in nature, but it is a nonterminating, nonrepeating decimal, and can never be exactly written down in that form.

Rounded off to 10 significant figures, and then progressively on down to nine, eight, seven, six, five, four, and three significant figures, π turns out to have values as follows:

$$3.141592654$$
$$3.14159265$$
$$3.1415927$$
$$3.141593$$
$$3.14159$$
$$3.1416$$
$$3.142$$
$$3.14$$

We can use as many significant figures we need when we encounter a constant of this type in a formula. (Another example is e, the natural logarithm base.)

ADDITION AND SUBTRACTION

When measured quantities are added or subtracted, determining the number of significant figures can involve subjective judgment. One way to resolve this is to expand all the values out to their plain decimal form (if possible), make the calculation, and then, at the end of the process, decide how many significant figures you can reasonably claim.

In some cases, the outcome of determining significant figures in a sum or difference is similar to what happens with multiplication or division. Take, for example, the sum $x + y$, where $x = 3.778800 \times 10^{-6}$ and $y = 9.22 \times 10^{-7}$. This calculation proceeds as follows:

$$x = 0.000003778800$$

$$y = 0.000000922$$

$$x + y = 0.0000047008$$
$$= 4.7008 \times 10^{-6}$$
$$= 4.70 \times 10^{-6}$$

In other instances, one of the values in a sum or difference is insignificant with respect to the other. Suppose that $x = 3.778800 \times 10^4$, while $y = 9.22 \times 10^{-7}$. The process of finding the sum goes like this:

$$x = 37,788$$

$$y = 0.000000922$$

$$x + y = 37,788.000000922$$
$$= 3.7788000000922 \times 10^4$$

In this case, y is so much smaller than x that it does not significantly affect the value of the sum. We can conclude that the sum here is the same as the larger number:

$$x + y = 3.7788 \times 10^4$$

 PROBLEM 3-6
What is the product of 1.001×10^5 and 9.9×10^{-6}, taking significant figures into account?

SOLUTION 3-6
Multiply the coefficients and the powers of 10 separately:

$$(1.001 \times 10^5)(9.9 \times 10^{-6}) = (1.001 \times 9.9) \times (10^5 \times 10^{-6})$$
$$= 9.9099 \times 10^{-1}$$
$$= 0.99099$$

We must round this to two significant figures, because that is the most we can legitimately claim. This particular expression does not have to be written out in power-of-10 form, because the exponent is within the range ±2 inclusive. Therefore:

$$(1.001 \times 10^5)(9.9 \times 10^{-6}) = 0.99$$

Quick Practice

Here are some practice problems that cover the material presented in this chapter. Solutions follow problems.

PROBLEMS

1. Write down the number 238,200,000,000,000 in scientific notation.

2. Write down the number 0.00000000678 in scientific notation.

3. Draw a number line that spans 4 orders of magnitude, from 1 to 10^4.

4. Draw a coordinate system with a horizontal scale that spans 2 orders of magnitude, from 1 to 100, and a vertical scale that spans 4 orders of magnitude, from 0.01 to 100.

5. What does 3.5562E+99 represent? How does it differ from 3.5562E–99?

SOLUTIONS

1. This is the equivalent of 2.382 multiplied by 100,000,000,000,000 (or 10^{14}), so in scientific notation, it is written as 2.382×10^{14}.

2. This is the equivalent of 6.78 multiplied by 0.000000001 (or 10^{-9}), so in scientific notation, it is written as 6.78×10^{-9}.

3. See Fig. 3-2.

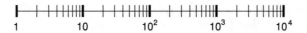

Fig. 3-2. Illustration for Quick Practice Problem and Solution 3.

4. See Fig. 3-3.

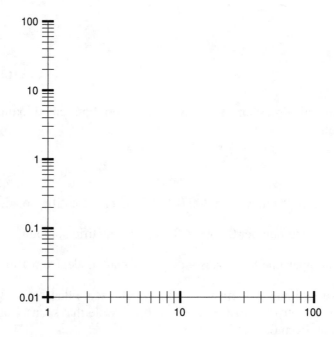

Fig. 3-3. Illustration for Quick Practice Problem and Solution 4.

5. The first expression represents 3.5562×10^{99}, which is a huge positive integer. The letter E means "times 10 to the power of." The digits that follow the E compose the exponent. The second expression represents 3.5562×10^{-99}. This is an extremely small positive rational number.

Quiz

This is an "open book" quiz. You may refer to the text in this chapter. A good score is 8 correct. Answers are in the back of the book.

1. Consider two numbers p and q, such that $p = q + 10{,}000$. By how many orders of magnitude do p and q differ?
    (a) 10,000.
    (b) 100.
    (c) 4.
    (d) More information is necessary to answer this.

2. How many significant figures does the numeric expression 3,700.00 have?
    (a) 6.
    (b) 4.
    (c) 3.
    (d) 2.

3. In the coordinate system shown by Fig. 3-4, the horizontal scale
    (a) is linear.
    (b) spans 1 order of magnitude.
    (c) spans 10 orders of magnitude.
    (d) spans 100 orders of magnitude.

4. In the coordinate system shown in Fig. 3-4, the largest value that can be plotted on the y axis, as it is drawn, differs from the smallest value that can be plotted on that same axis, as it is drawn, by a factor
    (a) that can't be determined without more information.
    (b) of 10^2.
    (c) of 10^3.
    (d) of 10^4.

5. Suppose you have a positive real number. Call it x. First, you square x. Then, you square the result of that. Next, you cube the result of that. Finally, you take the fourth power of the result of that. What is the final value in terms of x?
    (a) 1
    (b) x^{10}
    (c) x^{48}
    (d) There is no way to tell without knowing the exact value of x.

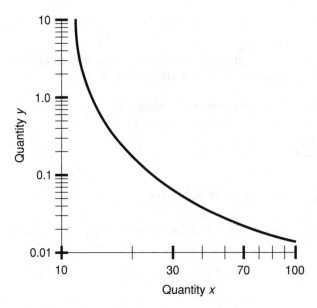

Fig. 3-4. Illustration for Quiz Questions 3 and 4.

6. Suppose you have a positive real number. Call it y. First, you square y. Then, you square the result of that. Next, you cube the result of that. Finally, you take the $\frac{1}{3}$ power of the result of that. What is the final value in terms of y?

    (a) 1
    (b) y^2
    (c) y^4
    (d) There is no way to tell without knowing the exact value of y.

7. Consider two numbers p and q, such that $p = q/100$. By how many orders of magnitude do p and q differ?

    (a) 100
    (b) 10
    (c) 2
    (d) None of the above

8. Suppose you have a positive real number, expressed to 4 significant figures. Call it z. First, you take the fourth power of z. Then, you divide the result by the square root of z. Finally, you square the result of that. When

you get the final answer, how many significant figures can you legitimately claim?

(a) 8.

(b) 6.

(c) 4.

(d) The value is theoretically exact, so you can claim as many significant figures as you want.

9. Suppose you have a positive real number, expressed to 4 significant figures. Call it w. First, you take the fourth power of w. Then, you divide the result by the square root of w. Finally, you take the zeroth power of that. When you get the final answer, how many significant figures can you legitimately claim?

(a) 8.

(b) 6.

(c) 4.

(d) The value is theoretically exact, so you can claim as many significant figures as you want.

10. Imagine a variable x that starts out as a large number negatively (for example, $x = -1000$), and increases in value, passing through $x = -100$, then $x = -10$, then $x = -1$, then $x = -0.1$, then $x = -0.01$, and so on, closer and closer to 0. Suppose that the value of x approaches 0, and gets arbitrarily close without limit, yet never reaches 0 or becomes positive. If x is expressed in scientific notation, how does the power of 10 change as x varies in this way?

(a) It gets larger and larger, and is always positive.

(b) It gets smaller and smaller, and is always negative.

(c) It starts out positive, gets smaller and smaller, passes through 0, and then becomes larger and larger negatively.

(d) It starts out negative, gets larger and larger (that is, smaller and smaller negatively), passes through 0, and then becomes larger and larger positively.

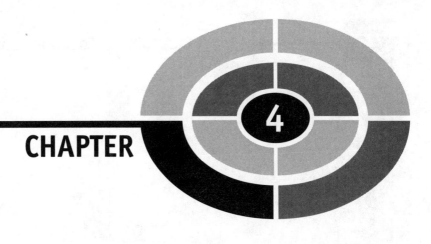

Coordinates in Two Dimensions

This chapter deals with coordinate systems and graphs in two dimensions (2D). You will often see these in scientific and engineering papers, and they are the sorts of graphs you'll encounter in technical math courses. They're also common in subjects such as economics, meteorology, geology, physics, electronics, and chemistry.

Cartesian Coordinates

The *Cartesian plane*, also called the *rectangular coordinate plane* or a system of *rectangular coordinates*, is defined by two number lines. Figure 4-1 illustrates the simplest possible set of rectangular coordinates. The number lines intersect at their zero points, and are perpendicular to each other. The "horizontal"

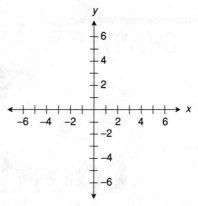

Fig. 4-1. The Cartesian plane is defined by two number lines that intersect at right angles.

(right/left) number line is called the *x axis*, and the "vertical" (up/down) number line is called the *y axis*.

ORDERED PAIRS AS POINTS

Figure 4-2 shows two specific points *P* and *Q*, plotted on the Cartesian plane. The coordinates of *P* are (–5, –4), and the coordinates of *Q* are (3,5). Any given

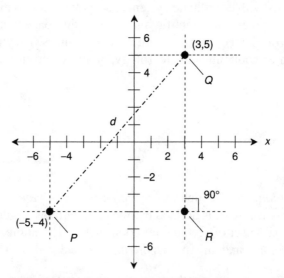

Fig. 4-2. Two points *P* and *Q*, plotted in rectangular coordinates, and a third point *R*, useful in calculating the distance *d* between *P* and *Q*.

point on the plane can be denoted as an *ordered pair* in the form (x,y), determined by the numerical values at which perpendiculars from the point intersect the x and y axes. In Fig. 4-2, the perpendiculars are shown as "vertical" and "horizontal" dashed lines. (Note that in the denotation of an ordered pair using parentheses and a comma, there is no space after the comma.)

ABSCISSA, ORDINATE, AND ORIGIN

In any graphing scheme, there is at least one *independent variable* and at least one *dependent variable*. The independent-variable coordinate (usually x) of a point on the Cartesian plane is known as the *abscissa* of the point, and the dependent-variable coordinate (usually y) is known as the *ordinate* of the point. The point $(0,0)$ is called the *origin* of the coordinate plane. In Fig. 4-2, point P has an abscissa of -5 and an ordinate of -4, and point Q has an abscissa of 3 and an ordinate of 5.

DISTANCE BETWEEN POINTS

Consider two different points $P = (x_0,y_0)$ and $Q = (x_1,y_1)$ on the Cartesian plane. The distance d between these two points can be found by determining the length of the *hypotenuse*, or longest side, of a right triangle PQR, where point R is the intersection of a "horizontal" line through P and a "vertical" line through Q. (In this case, "horizontal" means "parallel to the x axis," and "vertical" means "parallel to the y axis.") An example is shown in Fig. 4-2. Alternatively, we can use a "horizontal" line through Q and a "vertical" line through P, and consider their mutual intersection point.

The *Theorem of Pythagoras* (also known as the *Pythagorean theorem*) from plane geometry states that the square of the length of the hypotenuse of a right triangle is equal to the sum of the squares of the other two sides. In this case, that means:

$$d^2 = (x_1 - x_0)^2 + (y_1 - y_0)^2$$

and therefore:

$$d = [(x_1 - x_0)^2 + (y_1 - y_0)^2]^{1/2}$$

where the 1/2 power is the positive square root. In the situation shown in Fig. 4-2, the distance d between points $P = (x_0,y_0) = (-5,-4)$ and $Q = (x_1,y_1) = (3,5)$ is:

$$d = \{[3 - (-5)]^2 + [5 - (-4)]^2\}^{1/2}$$
$$= [(3 + 5)^2 + (5 + 4)^2]^{1/2}$$
$$= (8^2 + 9^2)^{1/2}$$
$$= (64 + 81)^{1/2}$$
$$= 145^{1/2}$$
$$= 12.04$$

This is accurate to four significant figures, as determined using a standard digital calculator that can find square roots.

PROBLEM 4-1
What is the distance between the two points (0,5) and (−3,−3) in Cartesian coordinates? Express the answer to four significant figures.

SOLUTION 4-1
Use the distance formula. Let $(x_0, y_0) = (0,5)$ and $(x_1, y_1) = (-3,-3)$. Then:

$$d = [(x_1 - x_0)^2 + (y_1 - y_0)^2]^{1/2}$$
$$= [(-3 - 0)^2 + (-3 - 5)^2]^{1/2}$$
$$= [(-3)^2 + (-8)^2]^{1/2}$$
$$= (9 + 64)^{1/2}$$
$$= 73^{1/2}$$
$$= 8.544$$

Simple Cartesian Graphs

Straight lines on the Cartesian plane are represented by *linear equations*. There are several forms in which a linear equation can be written. All linear equations can be reduced to a form where neither x nor y is raised to any power other than 0 or 1.

STANDARD FORM OF LINEAR EQUATION

The *standard form of a linear equation* in variables x and y consists of constant multiples of the two variables, plus another constant, all summed up to equal zero:

$$ax + by + c = 0$$

In this equation, the constants are a, b, and c. If a constant happens to be equal to 0, then it is not written down, nor is its multiple (by either x or y) written down. Here are some examples of linear equations in standard form:

$$2x + 5y - 3 = 0$$
$$5y - 3 = 0$$
$$2x - 3 = 0$$
$$2x = 0$$
$$5y = 0$$

The last two of these equations can be simplified to $x = 0$ and $y = 0$, by dividing each side by 2 and 5, respectively.

SLOPE OF A LINE

The *slope* of a straight line (often symbolized m) in the Cartesian xy-plane is defined as the ratio of the change in y (symbolized Δy) to the change in x (symbolized Δx) between any two distinct points on the line:

$$m = \Delta y / \Delta x$$

"Horizontal" lines (those parallel to the x axis) have slopes equal to 0. Lines that "ramp upward to the right" have positive slope. Lines that "ramp downward to the right" have negative slope. "Vertical" lines (those parallel to the y axis) have undefined slope.

SLOPE-INTERCEPT FORM OF LINEAR EQUATION

A linear equation in variables x and y can be manipulated so it is in a form that is easy to plot on the Cartesian plane. Here is how a linear equation in standard form can be converted to *slope-intercept form*:

$$ax + by + c = 0$$
$$ax + by = -c$$
$$by = -ax - c$$
$$y = (-a/b)x - c/b$$
$$y = (-a/b)x + (-c/b)$$

where a, b, and c are real-number constants, and $b \neq 0$. The quantity $-a/b$ is the slope of the line, an indicator of how steeply and in what sense the line slants. The quantity $-c/b$ represents the ordinate (or y-value) of the point at which the line crosses the y axis, where $x = 0$. This point is called the *y-intercept*. Let dx represent some change in the value of x on such a graph. Let dy represent the change in the value of y that results from this change in x. The ratio dy/dx is the slope of the line, and is symbolized m. Let k represent the y-intercept. Then m and k can be derived from the coefficients a, b, and c as follows, provided $b \neq 0$:

$$m = -a/b$$
$$k = -c/b$$

The linear equation can be rewritten in slope-intercept form as:

$$y = mx + k$$

When you want to plot the graph of a linear equation that appears in slope-intercept form in Cartesian coordinates, proceed as follows:

- Plot the point on the y axis where $y = k$.
- Move horizontally to the right by some whole number of units (call this number n).
- If the slope m is positive, move vertically upward by mn units.
- If the slope m is negative, move vertically downward by $|mn|$ units, where $|mn|$ represents the absolute value of the product mn.
- If $m = 0$, don't move up or down at all.
- Plot the resulting point on the plane where $x = n$ and $y = mn + k$.
- Connect the two points with a straight line.

Figures 4-3A and 4-3B illustrate the following linear equations as graphed in slope-intercept form:

$$y = 5x - 3$$
$$y = -x + 2$$

A positive slope indicates that the line ramps upward as you move from left to right, and a negative slope indicates that the line ramps downward as you move from left to right. A slope of 0 indicates a horizontal line. The slope of a vertical line is undefined because, in the form shown here, it requires that m be defined as a quotient in which the denominator is equal to 0.

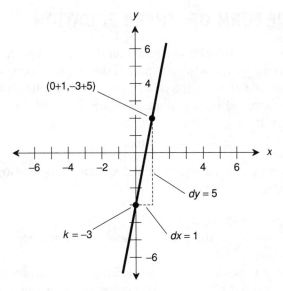

Fig. 4-3A. Graph of the linear equation $y = 5x - 3$.

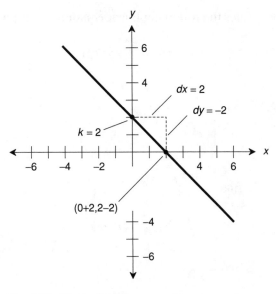

Fig. 4-3B. Graph of the linear equation $y = -x + 2$.

POINT-SLOPE FORM OF LINEAR EQUATION

It is difficult to plot a graph of a line based on the y-intercept (the point at which the line intersects the y axis) when the part of the graph of interest is far from the y axis. In this sort of situation, the *point-slope form* of a linear equation can be used. This form is based on the slope m of the line and the coordinates of a known point (x_0, y_0):

$$y - y_0 = m(x - x_0)$$

To plot a graph of a linear equation that appears in point-slope form, follow these steps in order:

- Plot the known point (x_0, y_0).
- Move horizontally to the right by some whole number of units (call this number n).
- If the slope m is positive, move vertically upward mn units.
- If the slope m is negative, move vertically downward $|mn|$ units, where $|mn|$ represents the absolute value of mn.
- If $m = 0$, don't move up or down at all.
- Plot the resulting point (x_1, y_1).
- Connect the points (x_0, y_0) and (x_1, y_1) with a straight line.

Figure 4-4A illustrates the following linear equation as graphed in point-slope form:

$$y - 104 = 3(x - 72)$$

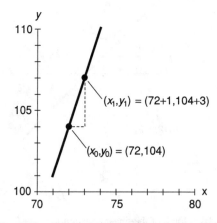

Fig. 4-4A. Graph of the linear equation $y - 104 = 3(x - 72)$.

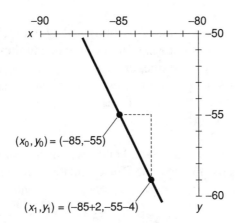

Fig. 4-4B. Graph of the linear equation $y + 55 = -2(x + 85)$.

Figure 4-4B is another graph of a linear equation in point-slope form:

$$y + 55 = -2(x + 85)$$

FINDING LINEAR EQUATION BASED ON GRAPH

Suppose we know the exact coordinates of two points P and Q in the Cartesian plane. These two points define a unique straight line L. Give the coordinates of the points these names:

$$P = (x_p, y_p)$$
$$Q = (x_q, y_q)$$

The slope m of line L can be found using either of the following formulas:

$$m = (y_q - y_p) / (x_q - x_p)$$
$$m = (y_p - y_q) / (x_p - x_q)$$

provided x_p is not equal to x_q. The point-slope equation of L can be determined based on the known coordinates of either point P or point Q. Therefore, either of the following formulas represent the line L:

$$y - y_p = m(x - x_p)$$
$$y - y_q = m(x - x_q)$$

PROBLEM 4-2

Consider the two points $P = (4,8)$ and $Q = (-1,2)$ on the Cartesian plane. What is the slope m of a line connecting these two points? What is the equation of that line, expressed in point-slope form? Remember that in an ordered pair for Cartesian coordinates, the first value is the x value, and the second value is the y value.

SOLUTION 4-2

To calculate m, we can use either of two formulas for the slope as given above. Let's use the first one. The calculation proceeds as follows, based on the input numbers $x_p = 4$, $y_p = 8$, $x_q = -1$, and $y_q = 2$:

$$m = (y_q - y_p)/(x_q - x_p)$$
$$= (2 - 8)/(-1 - 4)$$
$$= (-6)/(-5)$$
$$= 6/5$$

To determine the point-slope form of the equation for the line connecting the points, we can, again, use either of two general formulas given above. Let's use the first one, which is based on point P:

$$y - y_p = m(x - x_p)$$
$$y - 8 = (6/5)(x - 4)$$

Polar Coordinates

Two versions of the *polar coordinate plane* are shown in Figs. 4-5A and 4-5B. These systems are sometimes called *mathematician's polar coordinates*. One variable is plotted as an angle θ relative to a reference axis pointing to the right (or "east"), and the other variable is plotted as a distance (called the *radius*) r from the center. A coordinate point is thus denoted in the form of an ordered pair (θ,r). Some texts reverse the ordered pair, so coordinates are denoted in the form (r,θ). The (θ,r) notation is used in this book, because it is intuitively easier for most people to consider the angle as an independent variable (customarily listed first in ordered pairs) rather than as a dependent variable (customarily listed second). The point $(\theta,r) = (0,0)$ is called the *origin*.

RADIUS

In any polar plane, the radius coordinates are illustrated as concentric circles. The larger the circle, the greater the value of r. In Figs. 4-5A and 4-5B, the circles are

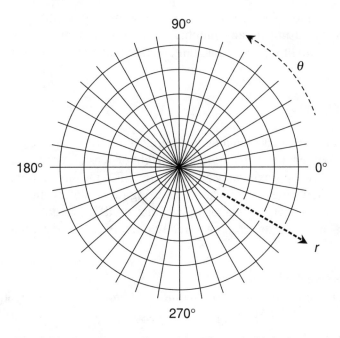

Fig. 4-5A. The polar coordinate plane. The angle θ is in degrees, and the radius r is in uniform increments.

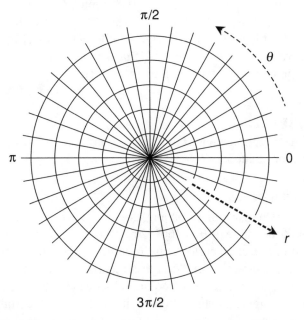

Fig. 4-5B. Another form of the polar coordinate plane. The angle θ is in radians, and the radius r is in uniform increments.

not labeled in units. You can do that. Imagine each concentric circle, working outward, as increasing by any number of units you want. For example, each radial division might represent one unit, or five units, or 10, or 100.

DIRECTION

Direction can be expressed in angular *degrees* (°) or *radians* (rad) counterclockwise from a reference axis pointing to the right or "east." A radian is the equivalent of $(180/\pi)$°, or approximately 57.3°. In Fig. 4-5A, the direction θ is in degrees. Figure 4-5B shows the same polar plane, using radians to express the direction. Regardless of whether degrees or radians are used, the angular scale is linear. The physical angle on the graph is directly proportional to the value of θ.

PROBLEM 4-3

Can the radius r in a polar-coordinate ordered pair be negative?

SOLUTION 4-3

Yes. In polar coordinates, it is all right to have a negative radius. If some point is specified with $r < 0$, we multiply r by –1 so it becomes positive, and then add or subtract 180° (π rad) to or from the direction angle. That is like saying "Drive 10 kilometers east" instead of "Drive negative 10 kilometers west." Negative radii must be allowed in order to graph certain equations in their entirety.

PROBLEM 4-4

Can the direction angle θ in a polar-coordinate ordered pair be less than 0° or more than 360°?

SOLUTION 4-4

Yes. Nonstandard direction angles are sometimes used in polar coordinates. If the value of θ is 360° (2π rad) or more, it represents more than one complete counterclockwise revolution from the 0° (0 rad) reference axis. If the direction angle is less than 0° (0 rad), it represents clockwise revolution instead of counterclockwise revolution. Nonstandard direction angles must be allowed in order to graph certain equations in their entirety.

Simple Polar Graphs

The graphs of some equations can be expressed more simply in polar coordinates than in Cartesian coordinates when the direction θ is expressed in radians. In the examples that follow, the "rad" abbreviation is eliminated, because it is understood that all angles are in radians.

CIRCLE CENTERED AT ORIGIN

The equation of a *circle centered at the origin* in the polar plane is given by the following formula:

$$r = a$$

where a is a real-number constant not equal to 0. The graph of this equation is illustrated in Fig. 4-6.

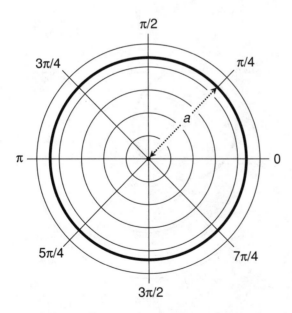

Fig. 4-6. Polar graph of a circle centered at the origin. The radius is equal to a.

CIRCLE PASSING THROUGH ORIGIN

The general form for the equation of a *circle passing through the origin* and centered at the point (θ_0, r_0) in the polar plane is as follows:

$$r = 2r_0 \cos(\theta - \theta_0)$$

where θ_0 and r_0 are real-number constants, and r_0 is not equal to 0. The abbreviation "cos" stands for the trigonometric cosine function, which can be determined for any known quantity using a scientific calculator. A generalized graph of the above equation is shown in Fig. 4-7.

ELLIPSE CENTERED AT ORIGIN

The equation of an *ellipse centered at the origin* in the polar plane is given by the following formula:

$$r = ab/(a^2 \sin^2 \theta + b^2 \cos^2 \theta)^{1/2}$$

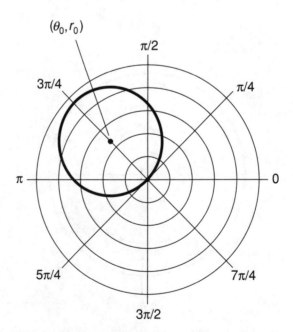

Fig. 4-7. Polar graph of a circle passing through the origin. The radius is equal to r_0.

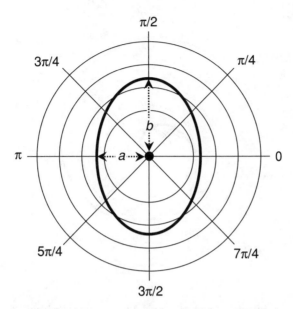

Fig. 4-8. Polar graph of an ellipse centered at the origin.
The values *a* and *b* represent the lengths of the semi-axes.

where *a* and *b* are real-number constants, neither of which is equal to 0. The abbreviations "sin" and "cos" stand for the trigonometric sine and cosine functions, which can be determined for any known quantities using a scientific calculator. A generalized graph of the above equation is shown in Fig. 4-8. In the figure, *a* represents the distance from the origin to the curve as defined along the "horizontal" radial axis, and *b* represents the distance from the origin to the curve as defined along the "vertical" radial axis. The values *a* and *b* represent the lengths of the *semi-axes* of the ellipse. The greater value is the length of the *major semi-axis*, and the lesser value is the length of the *minor semi-axis*.

SPIRAL

The general form of the equation of a *spiral* centered at the origin in the polar plane is given by the following formula:

$$r = a\theta$$

where *a* is a real-number constant not equal to 0. This constant determines the *pitch* (tightness) of the spiral. An example of this type of spiral, called the *spiral*

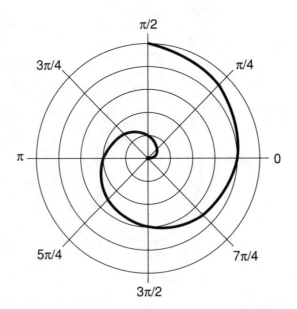

Fig. 4-9. Polar graph of a spiral. Illustration for Problem 4-5.

of Archimedes because of the uniform manner in which its radius increases as the angle increases, is illustrated in Fig. 4-9.

PROBLEM 4-5

What is the value of the constant, *a*, in the spiral shown in Fig. 4-9? What is the equation of this spiral? Assume that each radial division represents 1 unit.

SOLUTION 4-5

Note that if $\theta = \pi$, then $r = 2$. Therefore, we can solve for *a* by substituting this number pair in the general equation for the spiral. We know that $(\theta, r) = (\pi, 2)$, and that is all we need. Proceed like this:

$$r = a\theta$$
$$2 = a\pi$$
$$2/\pi = a$$

Therefore, $a = 2/\pi$, and the equation of the spiral is $r = (2/\pi)\theta$ or, in a somewhat simpler form without parentheses, $r = 2\theta/\pi$.

Navigator's Coordinates

Navigators and military people use a form of polar coordinate plane similar to that used by mathematicians and scientists. The radius is called the *range*, and real-world units are commonly specified, such as meters (m) or kilometers (km). The angle, or direction, is called the *azimuth*, *heading*, or *bearing*, and is measured in degrees clockwise from north. The basic scheme is shown in Fig. 4-10. The azimuth is symbolized α (the lowercase Greek alpha), and the range is symbolized r.

WHAT IS NORTH?

There are two ways of defining "north," or 0°. The more accurate, and thus the preferred and generally accepted, standard uses *geographic north*. This is the direction you would travel if you wanted to take the shortest possible route over the

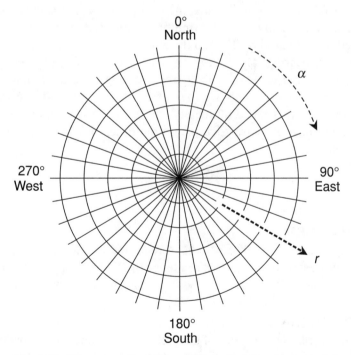

Fig. 4-10. The navigator's polar coordinate plane. The bearing α is in degrees, and the range r is usually in meters or kilometers.

earth's surface to the north geographic pole. In the northern hemisphere, geographic north corresponds almost exactly to a point on the horizon directly below *Polaris* (also called the *North Star*). The other standard uses *magnetic north*. This is the direction indicated by the needle in a magnetic compass.

For most locations on the earth's surface, there is a difference between geographic north and magnetic north. This difference, measured in degrees, is called the *declination*. Navigators in olden times had to know the declination for their location, when they couldn't use the stars to determine geographic north. Nowadays, there are electronic navigation systems such as the *Global Positioning System* (GPS) that make the magnetic compass irrelevant, provided the equipment is in working order.

STRICT RESTRICTIONS

In navigator's polar coordinates, the range can never be negative. Navigators will not talk about traveling –20 km on a heading of 270°, for example, when they really mean they are traveling 20 km on a heading of 90°. When working out certain problems, it's possible that the result will contain a negative range. If this happens, the value of r should be multiplied by –1, and the value of α should be increased or decreased by 180° so the result is at least 0° but less than 360°.

The azimuth, bearing, or heading must likewise conform to certain values. The smallest possible value of α is 0° (representing geographic north). As you turn clockwise as seen from above, the values of α increase through 90° (east), 180° (south), 270° (west), and ultimately approach 360° (north again).

We therefore have these restrictions on the ordered pair (α, r) in navigator's polar coordinates:

$$0 \leq \alpha < 360$$

$$r \geq 0$$

PROBLEM 4-6

Suppose you observe a radar screen and see a target (blip) directly southwest of your location, at a range of 10 km. What are the coordinates of this target? Express these coordinates as an ordered pair of the form (α, r).

SOLUTION 4-6

Given "real-world" units of kilometers for the range, we know that $r = 10$. Geographic southwest lies at a bearing of 225°. Therefore, $\alpha = 225$. The coordinates are:

$$(\alpha,r) = (225°,10)$$

Note that in navigator's polar coordinates, the angles are normally specified in degrees, not in radians.

Coordinate Conversions

In science and engineering, it is sometimes necessary to convert from one type of coordinate system to another. This can be done handily using computer programs designed for the purpose. Nevertheless, it is worthwhile to know the mathematical principles by which these programs work.

CARTESIAN TO MATHEMATICIAN'S POLAR

Figure 4-11 shows a point $P = (x_0, y_0) = (\theta_0, r_0)$ graphed on superimposed Cartesian and mathematician's polar coordinate systems. If we know the Cartesian coordinates, we can convert to mathematician's polar coordinates using these formulas:

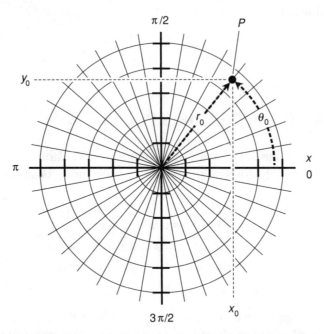

Fig. 4-11. Conversion between mathematician's polar and Cartesian (rectangular) coordinates. Each radial division represents 1 unit. Each division on the x and y axes represents 1 unit.

$$\theta_0 = \arctan (y_0/x_0) \text{ if } x_0 > 0$$
$$\theta_0 = 90° \text{ if } x_0 = 0 \text{ and } y_0 > 0 \text{ (for } \theta_0 \text{ in degrees)}$$
$$\theta_0 = 270° \text{ if } x_0 = 0 \text{ and } y_0 < 0 \text{ (for } \theta_0 \text{ in degrees)}$$
$$\theta_0 = \pi/2 \text{ if } x_0 = 0 \text{ and } y_0 > 0 \text{ (for } \theta_0 \text{ in radians)}$$
$$\theta_0 = 3\pi/2 \text{ if } x_0 = 0 \text{ and } y_0 < 0 \text{ (for } \theta_0 \text{ in radians)}$$
$$\theta_0 = 180° + \arctan (y_0/x_0) \text{ if } x_0 < 0 \text{ (for } \theta_0 \text{ in degrees)}$$
$$\theta_0 = \pi + \arctan (y_0/x_0) \text{ if } x_0 < 0 \text{ (for } \theta_0 \text{ in radians)}$$
$$r_0 = (x_0{}^2 + y_0{}^2)^{1/2}$$

If θ_0 turns out negative as determined according to any of the above formulas, add $360°$ or 2π rad to get the legitimate value, which should always be greater than or equal to $0°$ (0 rad) and less than $360°$ (2π rad).

MATHEMATICIAN'S POLAR TO CARTESIAN

Mathematician's polar coordinates can be converted to Cartesian coordinates using the following formulas:

$$x_0 = r_0 \cos \theta_0$$
$$y_0 = r_0 \sin \theta_0$$

CARTESIAN TO NAVIGATOR'S POLAR

In order to convert the coordinates of a point (x_0, y_0) in Cartesian coordinates to a point (α_0, r_0) in navigator's polar coordinates, use these formulas:

$$\alpha_0 = \arctan (x_0/y_0) \text{ if } y_0 > 0$$
$$\alpha_0 = 90° \text{ if } y_0 = 0 \text{ and } x_0 > 0$$
$$\alpha_0 = 270° \text{ if } y_0 = 0 \text{ and } x_0 < 0$$
$$\alpha_0 = 180° + \arctan (x_0/y_0) \text{ if } y_0 < 0$$
$$r_0 = (x_0{}^2 + y_0{}^2)^{1/2}$$

If α_0 turns out negative according to any of the above formulas, add $360°$ to get the legitimate value, which should always be greater than or equal to $0°$ but less than $360°$.

NAVIGATOR'S POLAR TO CARTESIAN

Here are the conversion formulas for translating a point (α_0, r_0) in navigator's polar coordinates to a point (x_0, y_0) in the Cartesian plane:

$$x_0 = r_0 \sin \alpha_0$$
$$y_0 = r_0 \cos \alpha_0$$

These are similar to the formulas you use to convert mathematician's polar coordinates to Cartesian coordinates, except the roles of the sine and cosine function are reversed.

MATHEMATICIAN'S POLAR TO NAVIGATOR'S POLAR

Sometimes you must convert from mathematician's polar coordinates to navigator's polar coordinates, or vice-versa. The radius or range of a particular point, r_0, is always the same in both systems, but the angles usually differ. If you know the direction angle θ_0 of a point in mathematician's polar coordinates and you want to find the equivalent azimuth α_0 in navigator's polar coordinates, first be sure θ_0 is expressed in degrees. Then use one of the following formulas, as applicable:

$$\alpha_0 = 90° - \theta_0 \text{ if } 0° \le \theta_0 \le 90°$$
$$\alpha_0 = 450° - \theta_0 \text{ if } 90° < \theta_0 < 360°$$

NAVIGATOR'S POLAR TO MATHEMATICIAN'S POLAR

If you know the azimuth α_0 in degrees of a distant point or target in navigator's polar coordinates, and you want to find the equivalent direction angle θ_0 in mathematician's polar coordinates, then you can use either of the following conversion formulas, depending on the value of α_0:

$$\theta_0 = 90° - \alpha_0 \text{ if } 0° \le \alpha_0 \le 90°$$
$$\theta_0 = 450° - \alpha_0 \text{ if } 90° < \alpha_0 < 360°$$

PROBLEM 4-7
Consider the point $(\theta_0, r_0) = (3\pi/4, 2)$ in mathematician's polar coordinates. What is the (x_0, y_0) representation of this point in Cartesian coordinates? Express the answer to four significant figures.

SOLUTION 4-7

Use the conversion formulas above:

$$x_0 = r_0 \cos \theta_0$$
$$y_0 = r_0 \sin \theta_0$$

Plugging in the numbers gives us these values, accurate to four significant figures:

$$x_0 = 2 \cos (3\pi/4) = 2 \times (-0.7071) = -1.414$$
$$y_0 = 2 \sin (3\pi/4) = 2 \times 0.7071 = 1.414$$

Thus, $(x_0, y_0) = (-1.414, 1.414)$.

PROBLEM 4-8

Suppose a radar set displaying navigator's polar coordinates indicates the presence of a hovering object at a bearing of 300° and a range of 40 km. If we say that a kilometer is the same as a "unit," what are the coordinates (θ_0, r_0) of this object in mathematician's polar coordinates? Express θ_0 in both degrees and radians.

SOLUTION 4-8

We are given coordinates $(\alpha_0, r_0) = (300°, 40)$. The value of r_0, the radius, is the same as the range, in this case 40 units. As for the angle θ_0, remember the conversion formulas given above. In this case, α_0 is greater than 90° and less than 360°. Therefore:

$$\theta_0 = 450° - \alpha_0$$
$$\theta_0 = 450° - 300° = 150°$$

This gives us the result $(\theta_0, r_0) = (150°, 40)$. To express θ_0 in radians, note that there are 2π radians in a full 360° circle, or π radians in a 180° angle. Also, note that 150° is exactly 5/6 of 180°. Therefore, $\theta_0 = 5\pi/6$ rad, and we can say that $(\theta_0, r_0) = (150°, 40) = (5\pi/6, 40)$.

Other Coordinate Systems

Here are a few other two-dimensional coordinate systems that are used in scientific and engineering applications.

LATITUDE AND LONGITUDE

Latitude and *longitude* coordinates uniquely define the positions of points on the surface of a sphere or in the sky. Even though a sphere is a three-dimensional (3D) object, its surface can be considered a 2D object that is curved or "warped." The same holds for apparent directions in the sky.

Figure 4-12A illustrates the latitude/longitude system commonly used to define locations on the earth's surface. Latitude "lines" are circles on the surface that run east and west, and that lie in planes parallel to the plane containing the equator. Longitude "lines" are circles on the surface that run north and south, each one intersecting the equator at a 90° angle, and all of them converging on the geographic poles. The *polar axis* connects to points at specifically chosen *antipodes* (opposite points) on the globe. These points are assigned latitudes $\theta = +90°$ (north geographic pole) and $\theta = -90°$ (south geographic pole). A *longitude reference axis* runs outward from the center of the sphere at a right angle

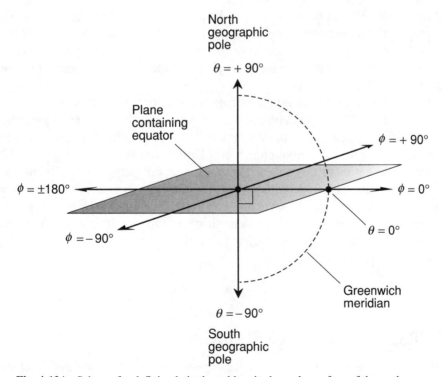

Fig. 4-12A. Scheme for defining latitude and longitude on the surface of the earth.

to the polar axis, passing through the equator, and it is assigned longitude $\phi = 0°$. Latitude angles θ are defined positively (north) and negatively (south) relative to the geometric plane containing the equator. Longitude angles ϕ are defined positively (east) and negatively (west) relative to the longitude reference axis. The angles are restricted as follows:

$$-90° \leq \theta \leq +90°$$
$$-180° < \phi \leq +180°$$

On the earth's surface, the half-circle connecting the 0° longitude line with the poles passes through Greenwich, England. It is known as the *Greenwich meridian* or the *prime meridian*. Longitude angles are defined with respect to this meridian. Negative angles are west of the prime meridian, and positive angles are east of the prime meridian.

CELESTIAL COORDINATES

Celestial latitude and *celestial longitude* are extensions of the earth's latitude and longitude into the heavens. Figure 4-12A, the same set of coordinates used for geographic latitude and longitude, applies to this system. An object whose celestial latitude and longitude coordinates are (θ,ϕ) appears at the zenith (directly overhead) in the sky from the point on the earth's surface whose latitude and longitude coordinates are (θ,ϕ).

Declination and *right ascension* define the positions of objects in the sky relative to the stars. (The term *declination* in celestial coordinates means something entirely different than the term *declination* that refers to the difference between geomagnetic north and geographic north. Don't confuse them!) Figure 4-12B illustrates the basics of this system. In celestial coordinates, the declination (θ) is identical to the celestial latitude. Right ascension (ϕ) is measured eastward from the *vernal equinox*, which is the position of the sun in the heavens at the moment spring begins in the northern hemisphere (on, or within a couple of days of, March 21). The angles are restricted as follows:

$$-90° \leq \theta \leq +90°$$
$$0° \leq \phi < 360°$$

HOURS, MINUTES, AND SECONDS

Astronomers use a peculiar scheme for expressing the angle of right ascension (RA). Instead of expressing it in degrees or radians, astronomers use *hours*,

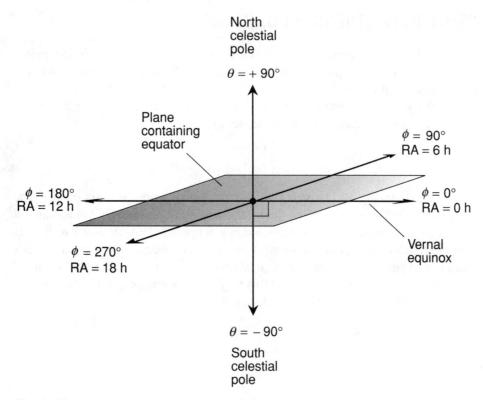

North
celestial
pole

$\theta = +90°$

Plane
containing
equator

$\phi = 90°$
RA = 6 h

$\phi = 180°$
RA = 12 h

$\phi = 0°$
RA = 0 h

$\phi = 270°$
RA = 18 h

Vernal
equinox

$\theta = -90°$

South
celestial
pole

Fig. 4-12B. Scheme for defining declination and right ascension in the sky.

minutes, and *seconds* based on 24 hours (h) in a complete circle (correspon-ding to the 24 hours in a day). That means each hour of right ascension is equi-valent to 15°. When RA angles are expressed as hours rather than as degrees, the following restriction applies for right ascension coordinates, also shown in Fig. 4-12B:

$$0\,h \leq RA < 24\,h$$

Minutes and seconds of right ascension, defined in this manner, turn out to be different from the fractional degree units by the same names. One minute of right ascension is 1/60 of an hour or 1/4 of an angular degree, and one second of right ascension is 1/60 of a minute or 1/240 of an angular degree. This com-pares with the pure angular definitions, where one minute of arc is 1/60 of a degree, and one second of arc is 1/60 of a minute or 1/3600 of a degree.

SEMILOG (*x*-LINEAR) COORDINATES

Figure 4-13 shows *semilogarithmic (semilog) coordinates* for defining points in a portion of the *xy*-plane. The independent-variable axis is linear, and the dependent-variable axis is logarithmic. The numerical values that can be depicted on the *y* axis are restricted to one sign or the other (positive or negative). Here, the graphable ranges of the variables are:

$$-1 \leq x \leq 1$$
$$0.1 \leq y \leq 10$$

The *y* axis in Fig. 4-13 spans two orders of magnitude (powers of 10). The span could be larger or smaller than this, but in any case the *y* values cannot extend to zero. Because the *x* axis covers a range all the way down to 0, it is meaningless to talk about the number of orders of magnitude it portrays. It's tempting to suggest that the *x* axis as shown in Fig. 4-13 covers an "infinite" number of orders of magnitude, but this might lead some people to imagine that the axis is infinitely long, which it clearly is not.

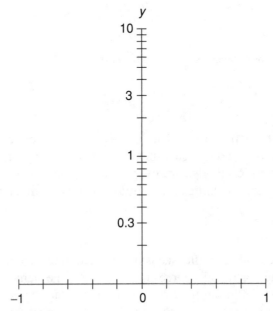

Fig. 4-13. The semilog *xy*-plane with linear *x* axis and logarithmic *y* axis.

SEMILOG (*y*-LINEAR) COORDINATES

Figure 4-14 shows semilog coordinates for defining points in a different portion of the *xy*-plane. In this case, the independent-variable axis is logarithmic, and the dependent-variable axis is linear. The numerical values that can be depicted on the *x* axis are restricted to one sign or the other (positive or negative). In this case, the graphable ranges of the variables are:

$$0.1 \le x \le 10$$
$$-1 \le y \le 1$$

The *x* axis in Fig. 4-14 spans two orders of magnitude (powers of 10). The span could be larger or smaller, but in any case the *x* values cannot extend to zero. The number of orders of magnitude covered by the *y* axis is undefined here, for the same reason the number of orders of magnitude for the *x* axis in Fig. 4-13 is undefined.

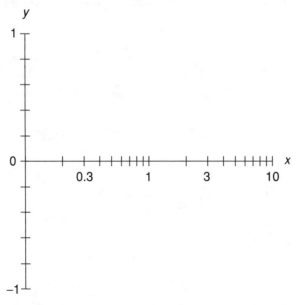

Fig. 4-14. The semilog *xy*-plane with logarithmic *x* axis and linear *y* axis.

LOG-LOG COORDINATES

Figure 4-15 shows *log-log coordinates* for defining points in a portion of the *xy*-plane. Both axes are logarithmic. The numerical values that can be depicted on either axis are restricted to one sign or the other (positive or negative). In this example, the graphable ranges of the variables are:

$$0.1 \leq x \leq 10$$

$$0.1 \leq y \leq 10$$

Both of the axes in Fig. 4-15 span two orders of magnitude (powers of 10). The span of either axis could be larger or smaller, but in any case the values cannot extend to zero.

PROBLEM 4-9

How do the celestial latitude and longitude of a star in the sky, such as Sirius (the "Dog Star"), change over the course of a 24-hour period?

SOLUTION 4-9

The celestial latitude of a star does not change over the course of a 24-hour period. The celestial longitude progresses from east to west, making

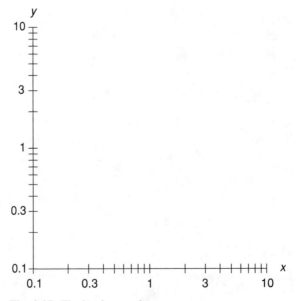

Fig. 4-15. The log-log *xy*-plane.

a complete circle every 24 hours. For example, if Sirius is at celestial longitude $\phi = 0°$ at 1:00 A.M. Eastern Standard Time (EST), then $\phi = -90°$ at 7:00 A.M. EST, $\phi = +180°$ at 1:00 P.M. EST, and $\phi = +90°$ at 7:00 P.M. EST. In the case of a star that is exactly at the north celestial pole or the south celestial pole, the celestial longitude is not defined. Polaris (the "North Star") is almost exactly at the north celestial pole.

PROBLEM 4-10

How do the declination and right ascension of a star in the sky, such as Sirius (the "Dog Star"), change over the course of a 24-hour period?

SOLUTION 4-10

Neither of these coordinates changes over the course of a 24-hour period, because declination and right ascension are defined with respect to the stars, not with respect to the surface of the earth. This is why astronomers prefer to specify declination and right ascension, rather than celestial latitude and longitude, for locating stars, nebulae, galaxies, and other distant objects in the heavens.

Quick Practice

Here are some practice problems that cover the material presented in this chapter. Solutions follow the problems.

PROBLEMS

1. Find the distance between the points (2,4) and (−3,12) on the Cartesian plane. Assume these coordinates are exact, and round the answer off to three significant figures.

2. Find the distance between the points (0,0) and (100,110) on the Cartesian plane. Assume these coordinates are exact, and round the answer off to five significant figures. Use scientific notation for all numerical values of 10,000 (10^4) or greater.

3. Suppose a line has a slope of 3 and passes through the point (0,−8) on the Cartesian plane. What is the equation of this line in slope-intercept form?

4. Suppose a line is parallel to the y axis, and runs through the point (2,0) on the Cartesian plane. What is the equation of this line? What is the slope of this line?

5. Suppose a line runs through the point (−4,7) and has a slope of −2 on the Cartesian plane. What is the equation of this line in slope-intercept form? In standard form?

SOLUTIONS

1. Let $(x_0,y_0) = (2,4)$, and let $(x_1,y_1) = (-3,12)$. The distance d between the points is:

$$d = [(x_1 - x_0)^2 + (y_1 - y_0)^2]^{1/2}$$
$$= [(-3 - 2)^2 + (12 - 4)^2)]^{1/2}$$
$$= [(-5)^2 + 8^2]^{1/2}$$
$$= (25 + 64)^{1/2}$$
$$= 89^{1/2} = 9.433981...$$

This rounds off to 9.43.

2. Let $(x_0,y_0) = (0,0)$, and let $(x_1,y_1) = (100,110)$. The distance d between the points is:

$$d = [(x_1 - x_0)^2 + (y_1 - y_0)^2]^{1/2}$$
$$= [(100 - 0)^2 + (110 - 0)^2]^{1/2}$$
$$= (100^2 + 110^2)^{1/2}$$
$$= (1.0000 \times 10^4 + 1.2100 \times 10^4)^{1/2}$$
$$= (2.2100 \times 10^4)^{1/2}$$
$$= 148.660687...$$

This rounds off to 148.66.

3. The slope-intercept form for a line is $y = mx + k$, where m is the slope and k is the y-intercept. We are given $m = 3$. The point (0,−8) is on the y axis because $x = 0$, so the y-intercept is −8. The equation is found by plugging in these numbers, as follows:

$$y = 3x + (-8)$$
$$y = 3x - 8$$

4. The equation of the line is simply $x = 2$. The slope is undefined, because the line is "vertical" (parallel to the y axis).

5. The point-slope form for a line is $y - y_0 = m(x - x_0)$, where m is the slope and (x_0, y_0) is a point on the line. We are given $(x_0, y_0) = (-4, 7)$ and $m = -2$. The equation in slope-intercept form is found by "plugging numbers in" and then simplifying, as follows:

$$y - y_0 = m(x - x_0)$$
$$y - 7 = -2[x - (-4)]$$
$$y - 7 = -2(x + 4)$$
$$y - 7 = -2x - 8$$
$$y = -2x - 1$$

This can be converted to standard form by adding the quantity $(2x + 1)$ to both sides, simplifying, and rearranging:

$$y + 2x + 1 = -2x - 1 + 2x + 1$$
$$2x + y + 1 = 0$$

Quiz

This is an "open book" quiz. You may refer to the text in this chapter. You may draw diagrams if that will help you visualize things. A good score is 8 correct. Answers are in the back of the book.

1. Consider the following equation:

$$r = 6 \cos (\theta - \pi/4)$$

What does the graph of this equation look like in polar coordinates?
(a) It is a straight line passing through the origin.
(b) It is a straight line that does not pass through the origin.
(c) It is a circle centered at the origin.
(d) It is a circle that is not centered at the origin.

2. Consider the following equation:

$$y - 3 = 7(x + 2)$$

Which of the following points lies on the graph of this equation in Cartesian coordinates?

(a) (2,3)
(b) (−2,3)
(c) (2,−3)
(d) (−2, −3)

3. In the graph of the equation $3x + y + 5 = 0$, what is the slope?

(a) 3
(b) −3
(c) 5
(d) −5

4. In the graph of the equation $y − 7 = 0$, what is the slope?

(a) 7
(b) −7
(c) 0
(d) It is undefined.

5. Consider two points P and Q on the Cartesian plane. Suppose that point P is located at the origin, and the coordinates of point Q are (x_q, y_q). Now consider a third point R with coordinates (x_r, y_r), such that $x_r = 2x_q$ and $y_r = 2y_q$. How does the distance d_{pq} between points P and Q compare with the distance d_{pr} between points P and R?

(a) There is no way to tell without more information.
(b) $d_{pr} = (d_{pq}^2 + d_{pr}^2)^{1/2}$
(c) $d_{pr} = 4d_{pq}$
(d) $d_{pr} = 2d_{pq}$

6. What is the distance d in mathematician's polar coordinates between the origin and the point $\theta = \pi/2$?

(a) $d = \pi/2$
(b) $d = \pi$
(c) $d = (\pi/2) = \pi^2/4$
(d) This is a meaningless question, because in polar coordinates, $\theta = \pi/2$ does not represent a point.

7. What is the distance d in mathematician's polar coordinates between the origin and the point $(\theta_0, r_0) = (−\pi/4, 3)$?

(a) $d = 3$

(b) $d = \pi/4$

(c) $d = 5$

(d) This is a meaningless question, because in polar coordinates, (θ_0, r_0) $= (-\pi/4, 3)$ does not represent a point.

8. Consider a town located halfway between the equator and the north geographic pole. What is the longitude of this town?

(a) $+45°$

(b) $-45°$

(c) $\pi/2$

(d) There is no way to tell without more information.

9. Suppose a radar set displaying navigator's polar coordinates indicates the presence of a hovering object directly northeast of us, and a range of 7 nautical miles. If we say that a nautical mile is the same as a "unit," what are the coordinates (θ_0, r_0) of this object in mathematician's (not navigators) polar coordinates, with θ_0 expressed in radians?

(a) $(\pi/4, 7)$

(b) $(\pi/2, 7)$

(c) $(3\pi/4, 7)$

(d) $(\pi, 7)$

10. Imagine a distant star that appears to pass, as the earth rotates, directly through the zenith point in the sky over a town located halfway between the equator and the north geographic pole. At a certain time (which varies over the course of a year) each day, the star appears at the zenith as observed from that town. What is the declination of this distant star?

(a) $+90°$

(b) $-90°$

(c) $+45°$

(d) There is no way to tell without more information.

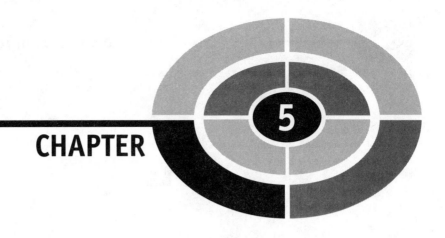

Coordinates in Three Dimensions

This chapter deals with three-dimensional (3D) coordinate systems and graphs. You will sometimes see these in scientific and engineering papers, and once in a while you'll encounter them in math courses, particularly in multivariable algebra, analysis, calculus, geometry, and topology.

Cartesian 3-Space

Figure 5-1 illustrates the simplest form of *Cartesian 3-space*, also called *rectangular 3D coordinates*. All three number lines have equal increments. (This is a perspective illustration, so the increments on the z axis appear distorted. A true 3D rendition would have the positive z axis perpendicular to the page.) The number lines intersect at a single point, corresponding to the zero points of each axis. Each of the axes is perpendicular to the other two.

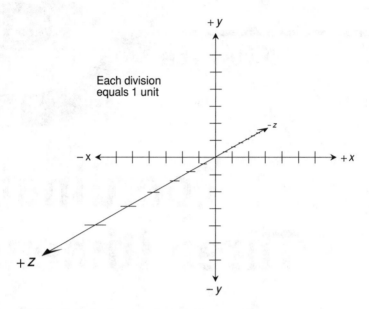

Fig. 5-1. Cartesian 3-space, also called *xyz*-space.

In most renditions of rectangular 3D coordinates, the positive *x* axis runs from the origin toward the viewer's right, and the negative *x* axis runs toward the left. The positive *y* axis runs upward, and the negative *y* axis runs downward. The positive *z* axis comes "out of the page," and the negative *z* axis extends "back behind the page." Sometimes the perspective is different (see the discussions of cylindrical coordinates below, for example), but the relative orientations of the positive and negative *x*, *y*, and *z* axes are standard.

ORDERED TRIPLES AS POINTS

Figure 5-2 shows two specific points, called *P* and *Q*, plotted in Cartesian 3-space. The coordinates of point *P* are (−5,−4,3), and the coordinates of point *Q* are (3,5,−2). Points are denoted as *ordered triples* in the form (*x*,*y*,*z*), where the first number represents the value on the *x* axis, the second number represents the value on the *y* axis, and the third number represents the value on the *z* axis. (When an ordered triple is written down, there are no spaces after the commas.)

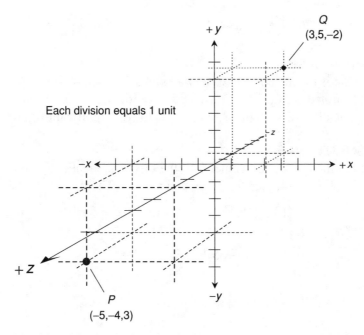

Fig. 5-2. Two points in Cartesian 3-space.

VARIABLES AND ORIGIN

In Cartesian 3-space, there are usually two independent-variable axes and one dependent-variable axis. The x and y axes represent independent variables; the z axis represents a dependent variable, the value of which depends on the x and y values.

In some scenarios, only one of the variables is independent, and two are dependent. In such cases, the independent variable is usually x. Rarely, you'll come across a situation in which none of the variables is dependent. This is the case, for example, when a correlation, but not a true mathematical relation, exists among the values of the variables. Plots of this sort typically appear as "swarms of points," representing the results of physical observations.

DISTANCE BETWEEN POINTS

Suppose there are two different points $P = (x_0, y_0, z_0)$ and $Q = (x_1, y_1, z_1)$ in Cartesian 3-space. The distance d between these two points can be found using this formula:

$$d = [(x_1 - x_0)^2 + (y_1 - y_0)^2 + (z_1 - z_0)^2]^{1/2}$$

PROBLEM 5-1

What is the distance between the points $P = (-5,-4,3)$ and $Q = (3,5,-2)$ illustrated in Fig. 5-2? Consider the coordinate values of P to be mathematically exact. Express the answer to five significant figures.

SOLUTION 5-1

Plug the coordinate values into the distance equation, where: $x_0 = -5$, $x_1 = 3$, $y_0 = -4$, $y_1 = 5$, $z_0 = 3$, and $z_1 = -2$. Then:

$$d = \{[3 - (-5)]^2 + [5 - (-4)]^2 + (-2 - 3)^2\}^{1/2}$$
$$= [8^2 + 9^2 + (-5)^2]^{1/2}$$
$$= (64 + 81 + 25)^{1/2}$$
$$= 170^{1/2}$$
$$= 13.038$$

PROBLEM 5-2

What is the distance between the origin and the point $R = (1,1,1)$ in Cartesian 3-space? Round the answer off to six significant figures, assuming the values of R are mathematically exact.

SOLUTION 5-2

Here, let $(x_0, y_0, z_0) = (0,0,0)$ and let $R = (x_1, y_1, z_1) = (1,1,1)$. Then, according to the distance formula, we obtain the following:

$$d = [(1 - 0)^2 + (1 - 0)^2 + (1 - 0)^2]^{1/2}$$
$$= (1^2 + 1^2 + 1^2)^{1/2}$$
$$= (1 + 1 + 1)^{1/2}$$
$$= 3^{1/2}$$
$$= 1.73205$$

Other 3D Coordinate Systems

Here are some alternative coordinate systems that are used in mathematics and science when working in 3D space.

CYLINDRICAL COORDINATES

Figures 5-3 and 5-4 show two systems of *cylindrical coordinates* for specifying the positions of points in space.

In the system shown in Fig. 5-3, we start with Cartesian 3-space. Then an angle θ is defined in the xy-plane, measured in degrees or radians (usually radians) counterclockwise from the positive x axis, which is called the *reference axis*. Given a point P in space, consider its *projection* point, P', on the xy-plane. The term *projection* in this context means that P' is directly "below" P, such that the line connecting these two points is parallel to the z axis. The position of P is defined by the ordered triple (θ,r,h). In this ordered triple, θ represents the angle in radians (from 0 to 2π) counterclockwise between P' and the positive x axis in the xy-plane, r represents the *radius* from P' to the origin, and h represents the *altitude* or *height* of P above the xy-plane. If h is negative, then P is below the xy-plane. This scheme for cylindrical coordinates is preferred by mathematicians, and also by some engineers and scientists.

In the system shown in Fig. 5-4, we again start with Cartesian 3-space. The xy-plane corresponds to the average surface of the earth (a horizontal plane) in the vicinity of the origin, and the z axis runs straight up (positive z values) and down (negative z values). The angle θ is defined in the xy-plane in degrees (from 0° to 360°) clockwise from the positive y axis, which corresponds to geographic

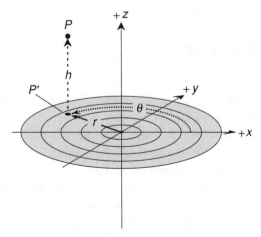

Fig. 5-3. Mathematician's form of cylindrical coordinates for 3D space.

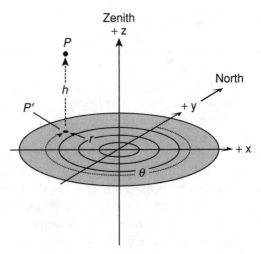

Fig. 5-4. Astronomer's and navigator's form of cylindrical coordinates for 3D space.

north. Given a point P in space, consider its projection P' onto the xy-plane. The position of P is defined by the ordered triple (θ,r,h), where θ represents the angle in degrees clockwise between P' and geographic north, r represents the radius from P' to the origin, and h represents the altitude of P above the xy-plane. If h is negative, then P is below the xy-plane (that will usually put P underground or underwater). This scheme is preferred by navigators and aviators.

SPHERICAL COORDINATES

Figures 5-5, 5-6, and 5-7 show three systems of *spherical coordinates* for defining points in space. The first two are used by astronomers and aerospace scientists, while the third one is of use to navigators and surveyors.

In the scheme shown in Fig. 5-5, the location of a point P is defined by the ordered triple (θ,ϕ,r) such that θ represents the declination of P, ϕ represents the right ascension of P, and r represents the distance or radius from P to the origin. In this example, angles are specified in degrees (except in the case of the astronomer's version of right ascension, which is expressed in hours, minutes, and seconds as defined in Chapter 4). Alternatively, the angles can be expressed in radians. This system is fixed relative to the stars.

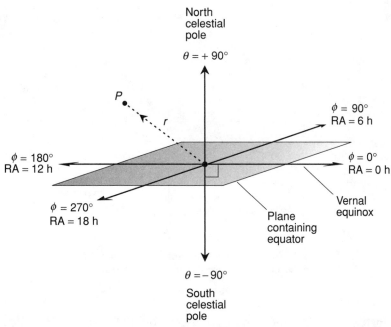

Fig. 5-5. Spherical coordinates for defining points in 3D space, where the angles represent declination (θ) and right ascension (ϕ).

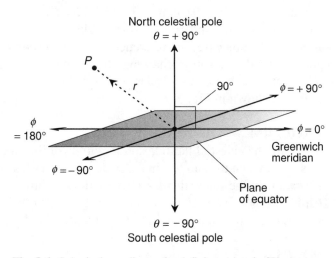

Fig. 5-6. Spherical coordinates for defining points in 3D space, where the angles represent celestial latitude (θ) and celestial longitude (ϕ).

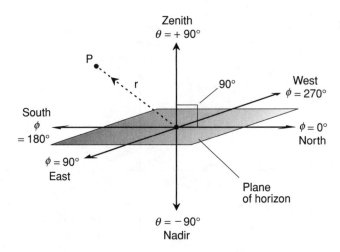

Fig. 5-7. Spherical coordinates for defining points in 3D space, where the angles represent elevation (θ) and azimuth (ϕ).

Instead of declination and right ascension, the variables θ and ϕ can represent celestial latitude and celestial longitude respectively, as shown in Fig. 5-6. This system is fixed relative to the earth, so it constantly rotates relative to the stars.

There's a third alternative: θ can represent the *elevation* (the angle above the horizon) and ϕ can represent the *azimuth* (the bearing or heading clockwise from geographic north). In this case, the reference plane corresponds to the horizon, not the equator, and the elevation can range between, and including, –90° (the nadir, or the point directly underfoot) and +90° (the zenith, or the point directly overhead). This is shown in Fig. 5-7.

PROBLEM 5-3
What are the celestial latitude and longitude of the sun on the first day of spring, when the sun lies in the plane of the earth's equator? What is the radius of the sun at this time? Write these coordinates as an ordered triple (θ,ϕ,r), where the angles are in degrees and the radius is in kilometers (km), for high noon in Greenwich, England.

SOLUTION 5-3
The celestial latitude (θ) of the sun on the first day of spring (March 21, the vernal equinox) is 0°, which is the same as the latitude of the

earth's equator. The celestial longitude (ϕ) of the sun depends on the time of day. It is 0° at high noon in Greenwich, England. From there, the celestial longitude of the sun proceeds west at the rate of 15° per hour (360° per 24 hours). The radius (r) of the sun is approximately 150,000,000 (1.5×10^8) km at all times. Thus, on the first day of spring at high noon in Greenwich:

$$(\theta,\phi,r) = [0°,0°,(150,000,000)]$$
$$= [0°,0°,(1.5 \times 10^8)]$$

PROBLEM 5-4

Suppose you stand in a huge, perfectly flat field and fly a kite on a string. The wind blows directly from the east. The point on the ground directly below the kite is 300 m away from you, and the kite is 400 m above the ground. If your body represents the origin and the units of a coordinate system are 1 m in size, what is the position of the kite in the cylindrical coordinate scheme preferred by navigators and aviators?

SOLUTION 5-4

The position of the kite is defined by the ordered triple (θ,r,h), where θ represents the angle measured clockwise from geographic north to a point directly under the kite, r represents the radius from a point on the ground directly under the kite to the origin, and h represents the altitude of the kite above the ground. Because the wind blows from the east, you know that the kite is directly west of the origin (represented by your body), so $\theta = 270°$. The value of r is the distance from your body to the point on the ground directly under the kite, which is given as 300 m, so $r = 300$. The kite is 400 m above the ground, so $h = 400$. Therefore, (θ,r,h) = (270°,300,400) in the system of cylindrical coordinates preferred by navigators and aviators.

Hyperspace

The Cartesian plane is defined by two number lines that intersect perpendicularly at their zero points. Cartesian 3-space is defined by three number lines that intersect at a single point, corresponding to the zero point of each line, and such that each line is perpendicular to the other two. Now let's see how Cartesian coordinates can work in *hyperspace*—any form of space having more than 3 dimensions.

IMAGINE THAT!

A system of rectangular coordinates in four dimensions, called *Cartesian 4-space*, is defined by four number lines that intersect at a single point, corresponding to the zero point of each line, and such that each of the lines is perpendicular to the other three. The lines form axes, representing variables such as w, x, y, and z. Alternatively, the axes can be labeled x_1, x_2, x_3, and x_4. Points are identified by *ordered quadruples* of the form (w,x,y,z) or (x_1,x_2,x_3,x_4). The origin is defined by $(0,0,0,0)$. As with the variables or numbers in ordered pairs and triples, there are no spaces after the commas when ordered quadruples are written down.

It can be tempting to draw an illustration such as Fig. 5-8 in an attempt to illustrate Cartesian 4-space. But when we start trying to plot points in this system, there is a problem. We can't define points in this rendition of 4D space without ambiguity. There are too many possible values of the ordered quadruple

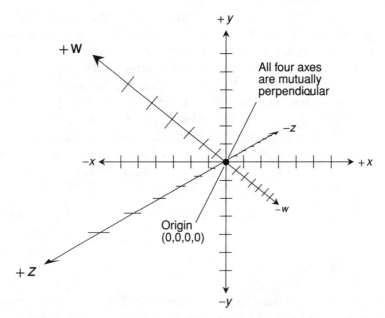

Fig. 5-8. An attempt to illustrate Cartesian 4-space. In a true 4D rectangular coordinate system, the w, x, y, and z axes are mutually perpendicular at the origin.

(w,x,y,z), and not enough points in 3D space to accommodate them all. In 3D space as we know it, four number lines such as those shown in Fig. 5-8 cannot be oriented so they intersect at a single point with all four lines perpendicular to the other three.

Mathematically, we can work with Cartesian 4-space, even though it cannot be directly visualized. This makes 4D geometry a powerful mathematical tool. As it turns out, the universe we live in requires four or more dimensions in order to be fully described. Albert Einstein was one of the first scientists to put forth the idea that a fourth dimension really exists.

TIME-SPACE

You've seen *time lines* in history books. You've seen them in graphs of various quantities, such as temperature, barometric pressure, or the price of a stock as a function of time. Isaac Newton, who developed classical physics, imagined time as flowing smoothly and unalterably. Time, according to classical or *Newtonian physics*, does not depend on space, nor space on time. Albert Einstein later showed that Newtonian physics is an oversimplification. But on a small scale, at moderate speeds, and over reasonable periods of time, Newtonian physics is an almost perfect system.

Wherever you are, however fast or slow you travel, and no matter what else you do, the cosmic clock, according to Newtonian physics, keeps ticking at the same absolute rate. In most practical scenarios, this model works quite well; its imperfections are not evident. Mathematically, we can envision a time line passing through 3D space, perpendicular to all three spatial axes such as the intersections between two walls and the floor of a room. The time axis passes through three-space at some chosen origin point.

In four-dimensional (4D) *Cartesian time-space* (or simply *time-space*), each point follows its own time line or curve. Assuming none of the points is in motion with respect to the origin, all the points follow straight time lines, and they are all perpendicular to 3D space. This sort of situation can be portrayed as shown in Fig. 5-9, with one spatial dimension removed, rendering 3D space as a flat geometric plane. Points that move at constant, but nonzero velocity (that is, at constant speed in an unchanging direction) with respect to the origin travel in straight lines; however these lines are not perpendicular to the 3D space. The greater the speed of a point relative to the origin, the sharper the angle of the path that it follows with respect to 3D space. Points that accelerate (move with changing speed or direction) follow curved paths through time-space.

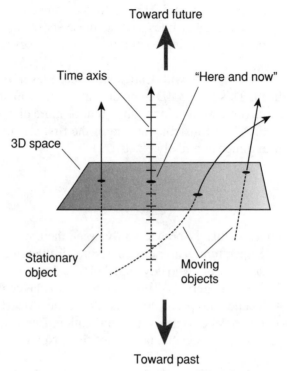

Fig. 5-9. Time-space can be shown in dimensionally reduced form by rendering 3D space as a plane, and adding a time axis perpendicular to that plane. Point-sized objects appear as straight or curved lines in this system.

POSITION VERSUS MOTION

Suppose we choose the center of the sun as the origin point for a Cartesian 3-space coordinate system. Imagine that the x and y axes lie in the plane of the earth's orbit around the sun. Suppose the positive x axis runs from the sun through the earth's position in space on March 21, and thence onward into deep space (roughly towards the constellation Virgo). The negative x axis runs from the sun through the earth's position on September 21 (roughly through the constellation Pisces). Suppose the positive y axis runs from the sun through the earth's position on June 21 (roughly toward the constellation Sagittarius), and the negative y axis runs from the sun through the earth's position on December 21 (roughly toward Gemini). Then the positive z axis runs from the sun toward the north celestial pole (in the direction of Polaris, the North Star), and the negative z axis runs from the sun toward the south celestial pole. Let each division on the coordinate axes represent one-quarter of an *astronomical unit* (AU), where 1 AU is defined as the mean distance of the earth from the sun (about 150,000,000 km or 93,000,000 mi). Figure 5-10 shows this coordinate system, with the earth on the positive x axis, at a distance of 1 AU. The coordinates of the earth at this time are (1,0,0) in the *xyz*-space we have defined. Each division on the axes in Fig. 5-10 represents 1/4 AU (37,500,000 km or about 23,300,000 mi).

Of course, the earth doesn't remain fixed; it orbits the sun. Let's take away the z axis in Fig. 5-10 and replace it with a time axis called t. The earth's path through this dimensionally-reduced time-space is not a straight line, but instead is a helix as shown in Fig. 5-11. The distance of the earth from the t axis remains nearly constant, although it varies a little because the earth's orbit around the sun is not a perfect circle. Every quarter of a year, the earth advances 90° around the helix. Theoretically, the earth accelerates as it travels around the sun, because its velocity (speed and direction) does not remain the same. The speed is nearly constant, but the direction changes. This is why the path of the earth through time-space is not a straight line.

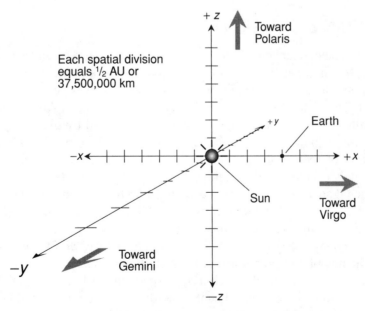

Fig. 5-10. A Cartesian coordinate system for the position of the earth in 3D space.

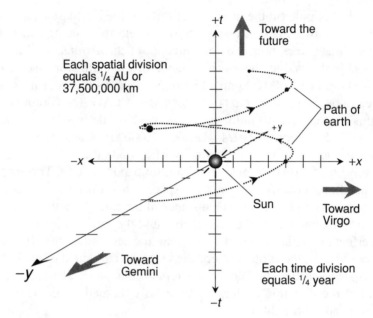

Fig. 5-11. A dimensionally reduced Cartesian system for rendering the path of the earth through 4D time-space.

CARTESIAN *n*-SPACE

A system of rectangular coordinates in five dimensions defines *Cartesian 5-space*. There are five number lines, all of which intersect at a point corresponding to the zero point of each line, and such that each of the lines is perpendicular to the other four. The resulting axes can be called v, w, x, y, and z. Alternatively they can be called x_1, x_2, x_3, x_4, and x_5. Points are identified by *ordered quintuples* such as (v,w,x,y,z) or (x_1,x_2,x_3,x_4,x_5). The origin is defined by $(0,0,0,0,0)$.

A system of rectangular coordinates in *Cartesian n-space* (where n is any positive integer) consists of n number lines, all of which intersect at their zero points, such that each of the lines is perpendicular to all the others. The axes can be named x_1, x_2, $x_3,\ldots$, and so on up to x_n. Points in Cartesian n-space can be uniquely defined by *ordered* n-*tuples* of the form $(x_1,x_2,x_3,\ldots,x_n)$.

There is nothing to stop us from dreaming up a *Cartesian 25-space* in which the coordinates of the points are ordered 25-tuples $(x_1,x_2,x_3,\ldots,x_{25})$, none of which are time. Alternatively, such a hyperspace might have 24 spatial dimensions and one time dimension. Then the coordinates of a point would be defined by the ordered 25-tuple $(x_1,x_2,x_3,\ldots,x_{24},t)$. In fact, there is no particular reason we can't have a Cartesian space with an infinite number of dimensions.

DISTANCE FORMULAS

In n-dimensional Cartesian space, the shortest distance between any two points can be found by means of a formula similar to the distance formulas for 2D and 3D space. The distance thus calculated represents the length of a straight line segment connecting the two points.

Suppose there are two points in Cartesian n-space, defined as follows:

$$P = (x_1, x_2, x_3, \ldots, x_n)$$
$$Q = (y_1, y_2, y_3, \ldots, y_n)$$

The length of the shortest possible path between points P and Q, written $|PQ|$, is equal to either of the following:

$$|PQ| = [(y_1 - x_1)^2 + (y_2 - x_2)^2 + (y_3 - x_3)^2 + \ldots + (y_n - x_n)^2]^{1/2}$$
$$|PQ| = [(x_1 - y_1)^2 + (x_2 - y_2)^2 + (x_3 - y_3)^2 + \ldots + (x_n - y_n)^2]^{1/2}$$

PROBLEM 5-5

Find the distance $|PQ|$ between the points $P = (4, -6, -3, 0)$ and $Q = (-3, 5, 0, 8)$ in Cartesian 4-space. Assume the coordinate values to be exact; express the answer to four significant figures.

SOLUTION 5-5

Assign the numbers in these ordered quadruples the following values: $x_1 = 4$, $x_2 = -6$, $x_3 = -3$, $x_4 = 0$, $y_1 = -3$, $y_2 = 5$, $y_3 = 0$, and $y_4 = 8$. Then plug these values into either of the above two distance formulas. Let's use the first formula:

$$|PQ| = \{(-3 - 4)^2 + [5 - (-6)]^2 + [0 - (-3)]^2 + (8 - 0)^2\}^{1/2}$$
$$= [(-7)^2 + 11^2 + 3^2 + 8^2]^{1/2}$$
$$= (49 + 121 + 9 + 64)^{1/2}$$
$$= 243^{1/2}$$
$$= 15.59$$

Quick Practice

Here are some practice problems that cover the material presented in this chapter. Solutions follow the problems.

PROBLEMS

1. What is the distance between the point $(x,y,z) = (6,7,8)$ and the origin in Cartesian 3-space? Assume these coordinates are exact, and round the answer off to three significant figures.

2. What is the distance between the points $(x_0,y_0,z_0) = (-4,-2,5)$ and $(x_1,y_1,z_1) = (1,4,-3)$ in Cartesian 3-space? Assume these coordinates are exact, and round the answer off to four significant figures.

3. Plot the point $(\theta,r,h) = (3\pi/4,6,8)$ in the cylindrical coordinate system commonly used by mathematicians.

4. What is the distance of the point $(\theta,r,h) = (3\pi/4,6,8)$ from the origin in the cylindrical coordinate system commonly used by mathematicians? Assume these coordinates are exact, and round the answer off to three significant figures.

5. Plot the point $(\theta,r,h) = (135°,3,4)$ in the cylindrical coordinate system commonly used by navigators and aviators. Assume these coordinates are exact.

SOLUTIONS

1. The origin is the point $(0,0,0)$. Using the distance formula for 3D space to find the distance d between the points $(0,0,0)$ and $(6,7,8)$, we calculate as follows:

$$d = [(6 - 0)^2 + (7 - 0)^2 + (8 - 0)^2]^{1/2}$$
$$= (6^2 + 7^2 + 8^2)^{1/2}$$
$$= (36 + 49 + 64)^{1/2}$$
$$= 149^{1/2}$$
$$= 12.2$$

2. Using the formula for 3D space to find the distance d between the points $(-4,-2,5)$ and $(1,4,-3)$, we calculate as follows:

$$d = \{[1 - (-4)]^2 + [4 - (-2)]^2 + (-3 - 5)^2\}^{1/2}$$
$$= [5^2 + 6^2 + (-8)^2]^{1/2}$$
$$= (25 + 36 + 64)^{1/2}$$
$$= 125^{1/2}$$
$$= 11.18$$

3. Refer to Fig. 5-12.

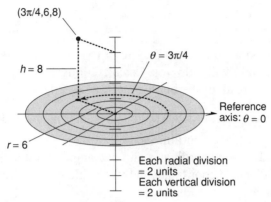

Fig. 5-12. Illustration for Quick Practice Problem and Solution 3.

4. This can be calculated by finding the length of the diagonal of a rectangle with sides of lengths r and h. This rectangle lies in a plane, so the 2D distance formula can be used, substituting r for x and h for y. We thus find the distance d between the points $(0,0)$ and $(6,8)$ in a Cartesian system, as follows:

$$d = [(6 - 0)^2 + (8 - 0)^2]^{1/2}$$
$$= (6^2 + 8^2)^{1/2}$$
$$= (36 + 64)^{1/2}$$
$$= 100^{1/2}$$
$$= 10$$

This is exact, but it can be expressed as 10.0.

5. Refer to Fig. 5-13.

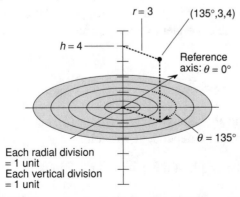

Fig. 5-13. Illustration for Quick Practice Problem and Solution 5.

Quiz

This is an "open book" quiz. You may refer to the text in this chapter. You may draw diagrams if that will help you visualize things. A good score is 8 correct. Answers are in the back of the book.

1. In a spherical 3D coordinate system, a point is uniquely defined according to
   - (a) three distance coordinates.
   - (b) two distance coordinates and one angle coordinate.
   - (c) one distance coordinate and two angle coordinates.
   - (d) three angle coordinates.

2. Consider a point P in Cartesian 3D space. Suppose the x, y, and z coordinates of P are each doubled. What will happen to the distance of P from the origin?
   - (a) It will increase by a factor of 2.
   - (b) It will increase by a factor of 2^2, or 4.
   - (c) It will increase by a factor of 2^3, or 8.
   - (d) It is impossible to answer this without more information.

3. Consider a point P in Cartesian 3D space. Suppose the x coordinate of P is doubled, but the y and z coordinates remain the same. What will happen to the distance of P from the origin?
   - (a) It will increase by a factor of 2.
   - (b) It will increase by a factor of 2^2, or 4.
   - (c) It will increase by a factor of 2^3, or 8.
   - (d) It is impossible to answer this without more information.

4. Imagine a Cartesian 3D coordinate system in which the axes are labeled according to variables x, y, and z. Suppose there are two independent variables and one dependent variable. The independent variables are usually denoted by
   - (a) the x and y axes.
   - (b) the x and z axes.
   - (c) the y and z axes.
   - (d) none of the above answers (a), (b), or (c), because there can never be more than one independent variable.

5. What would the graph of the equation $r = 5$ look like in a spherical coordinate system?

(a) A straight line.
(b) A circle.
(c) A sphere.
(d) It depends on whether the coordinate system is the one preferred by astronomers and aerospace scientists, or the one preferred by navigators and surveyors.

6. Consider two distant stars, called P and Q, whose positions in space are defined by an earth-centered spherical coordinate system where θ denotes the celestial latitude, ϕ denotes the celestial longitude, and r denotes the radius in light years. (A light year is the distance that a ray of visible light travels in one earth year—approximately 10^{13} km or 6×10^{12} mi.) Suppose that at a given moment in time, the coordinates of the two stars P and Q are:

$$P = (\theta_p, \phi_p, r_p) = (+30°, -20°, 48)$$
$$Q = (\theta_q, \phi_q, r_q) = (+15°, -10°, 96)$$

How do the distances of these two stars, as measured from the earth, compare?

(a) They are the same distance from the earth.
(b) Star P is half as far from the earth as star Q.
(c) Star P is twice as far from the earth as star Q.
(d) There is no way to compare the distances without more information.

7. Consider again the two distant stars, P and Q, described in Question 6. How do the values of the coordinates θ_p and θ_q change with the passage of time?

(a) The values of both coordinates cycle from 0° through –90°, 180°, +90°, and finally back to 0° again, once per day.
(b) The value of θ_p increases, while the value of θ_q decreases.
(c) The value of θ_p decreases, while the value of θ_q increases.
(d) They both remain constant with the passage of time.

8. In a cylindrical 3D coordinate system, a point is uniquely defined according to

(a) three distance coordinates.
(b) two distance coordinates and one angle coordinate.
(c) one distance coordinate and two angle coordinates.
(d) three angle coordinates.

9. Imagine Cartesian 4D time-space, consisting of conventional 3D space with time added as an extra dimension. Suppose the variable x represents kilometers (km) east and west from a defined origin point P, the variable y represents kilometers north and south from P, and the variable z represents kilometers above and below P. Suppose the time variable, t, is defined in seconds (s). Now imagine that you drive a car at a constant speed of 20 m/s along a straight highway running due north. The path along which you travel through Cartesian time-space can be represented by

(a) a sphere.
(b) a circle.
(c) a straight line.
(d) a curved line.

10. Imagine the same scenario as that of Question 9, except that in this case, rather than traveling at a constant speed along the highway running due north, you start from a dead stop and floor the gas pedal, thereby accelerating. The path along which you travel through Cartesian time-space can be represented by

(a) a sphere.
(b) a circle.
(c) a straight line.
(d) a curved line.

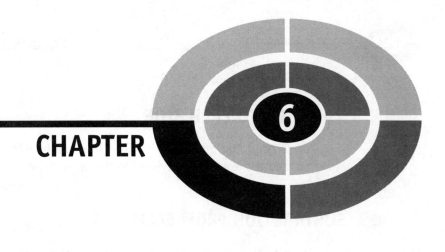

Equations in One Variable

The objective of solving a single-variable equation is to get it into a form where the expression on the left-hand side of the equals sign is simply the variable sought (such as x), and the expression on the right-hand side of the equals sign does not contain that variable.

Operational Rules

There are several ways in which an equation in one variable can be manipulated to obtain a solution, assuming a solution exists. Any and all of the principles of calculation outlined in Chapter 2 can be applied toward this result. In addition, the following rules can be applied in any order, and any number of times.

ADDITION TO EACH SIDE

Any defined constant, variable, or expression can be added to either side of an equation, and the result is equivalent to the original equation.

SUBTRACTION FROM EACH SIDE

Any defined constant, variable, or expression can be subtracted from either side of an equation, and the result is equivalent to the original equation.

MULTIPLICATION OF EACH SIDE

Both sides of an equation can be multiplied by a defined, nonzero constant, variable, or expression, and the result is equivalent to the original equation.

DIVISION OF EACH SIDE

Both sides of an equation can be divided by a nonzero constant, by a variable that cannot attain a value of zero, or by an expression that cannot attain a value of zero over the range of its variable(s), and the result is equivalent to the original equation.

BEWARE!

When both sides of an equation are divided by a variable, it is important to ensure that the variable cannot attain a value of zero. It is also not generally admissible to take both sides of an equation to a power (for example, to square both sides).

PROBLEM 6-1
Solve the equation $x + 45 = 33$ for x using one of the above principles.

SOLUTION 6-1
In this case, the equation can be solved by subtracting 45 from each side. This changes the expression on the left-hand side of the equality

symbol to x, and the expression on the right-hand side to -12. This is a direct statement of the solution: $x = -12$. Here is the process, broken down into every single step:

$$x + 45 = 33$$
$$(x + 45) - 45 = 33 - 45$$
$$x + (45 - 45) = -12$$
$$x + 0 = -12$$
$$x = -12$$

PROBLEM 6-2

Give an example of why it is not possible, in general, to take each side of an equation to a power and be certain of a valid result.

SOLUTION 6-2

Consider the equation $x^2 = 4$. In this case, x is a number such that, when squared, the result is equal to 4. Suppose we take the 1/2 power of each side of this equation in an attempt to solve it. The 1/2 power of a quantity is defined as the *positive square root* of that quantity. We get the following sequence of steps:

$$x^2 = 4$$
$$(x^2)^{1/2} = 4^{1/2}$$
$$x^{(2 \times 1/2)} = 2$$
$$x^1 = 2$$
$$x = 2$$

Each of these steps seems valid, but the solution thereby derived is incomplete. It overlooks the fact that x can also be equal to -2.

Linear Equations

A *linear equation in one variable* is an equation in which there is a single unknown, and that unknown is not raised to a power (other than the first power). Such an equation can always be reduced to the following standard form:

$$ax + b = 0$$

where a and b are constants, $a \neq 0$, and x is the unknown (that is, the variable).

GENERAL SOLUTION

A linear equation in a variable x, once it has been put into standard form, can be solved as follows:

$$ax + b = 0$$
$$ax + b - b = -b$$
$$ax = -b$$
$$x = -b/a$$

ALTERNATIVE FORMS

Here is another way a linear equation can appear. This time, let's use the variable y, and let's call the constants c and d, where $c \neq 0$. Then the following is a linear equation:

$$cy = d$$

Dividing both sides by c, we get this:

$$y = d/c$$

Here is another form in which a linear equation can appear:

$$kz + m = nz + p$$

where k, m, n, and p are constants such that $k \neq n$, and z is the unknown for which a solution is to be found. This equation can be solved as follows:

$$kz + m = nz + p$$
$$kz = nz + p - m$$
$$kz - nz = p - m$$
$$(k - n)z = p - m$$
$$z = (p - m)/(k - n)$$

PROBLEM 6-3
Solve the linear equation $4x - 5 = 2x + 7$.

SOLUTION 6-3
Here is how the equation is manipulated to solve for x:

$$4x - 5 = 2x + 7$$
$$(4x - 5) + 5 = (2x + 7) + 5$$
$$4x + (-5 + 5) = 2x + (7 + 5)$$
$$4x = 2x + 12$$
$$4x - 2x = 2x + 12 - 2x$$
$$2x = 12$$
$$x = 6$$

PROBLEM 6-4

Reduce the linear equation $4x - 5 = 2x + 7$ to standard form.

SOLUTION 6-4

Here is how the equation is manipulated to put it in standard form:

$$4x - 5 = 2x + 7$$
$$(4x - 5) - 7 = 2x$$
$$4x - 12 = 2x$$
$$4x - 12 - 2x = 0$$
$$2x - 12 = 0$$

PROBLEM 6-5

Solve the linear equation $2x - 12 = 0$, and see how much more quickly, in this situation, the process proceeds when the equation has been reduced to standard form.

SOLUTION 6-5

Here is how the equation, expressed in standard form, is manipulated to solve for x:

$$2x - 12 = 0$$
$$2x - 12 + 12 = 12$$
$$2x = 12$$
$$x = 6$$

Quadratic Equations

A *one-variable, second-order equation*, also called a *second-order equation in one variable* or, more often, a *quadratic equation*, can be written in the following standard form:

$$ax^2 + bx + c = 0$$

where a, b, and c are constants, $a \neq 0$, and x is the variable. The constants a, b, and c are also known as *coefficients*.

SOME EXAMPLES

Any equation that can be converted into the above form is a quadratic equation. Alternative forms, in which the variable is x, are:

$$mx^2 + nx = p$$
$$qx^2 = rx + s$$
$$(x + t)(x + u) = 0$$

where m, n, p, q, r, s, t, and u are constants. Here are two specific examples of quadratic equations that are not in standard form:

$$-3x^2 - 4x = 2$$
$$4x^2 = -3x + 5$$

GET IT INTO FORM

Some quadratic equations are easy to solve. Others are difficult. The first step in finding the value(s) of the variable in a quadratic equation is to get the equation into standard form.

The first equation above can be reduced to standard form by subtracting 2 from each side:

$$-3x^2 - 4x = 2$$
$$-3x^2 - 4x - 2 = 0$$

The second equation above can be reduced to standard form by adding $3x$ to each side and then subtracting 5 from each side:

$$4x^2 = -3x + 5$$
$$4x^2 + 3x = 5$$
$$4x^2 + 3x - 5 = 0$$

The *factored form* of a quadratic equation is convenient, because an equation denoted this way can be solved without having to do much work. Here is an example:

$$(x + 4)(x - 5) = 0$$

The expression on the left-hand side of the equals sign is zero if and only if either of the two factors is zero. If $x = -4$, then the equation becomes:

$$(-4 + 4)(-4 - 5) = 0 \times -9$$
$$= 0$$

If $x = 5$, then the equation becomes:

$$(5 + 4)(5 - 5) = 9 \times 0$$
$$= 0$$

Simply take the additive inverses (negatives) of the constants in each factor to get the solutions to a quadratic equation in factored form.

THE QUADRATIC FORMULA

Examine these two quadratic equations:

$$-3x^2 - 4x = 2$$
$$4x^2 = -3x + 5$$

These can be reduced to standard form:

$$-3x^2 - 4x - 2 = 0$$
$$4x^2 + 3x - 5 = 0$$

These equations are difficult to put into factored form. But there is a formula, known as the *quadratic formula*, that can be used to solve quadratic equations such as these.

Consider the following general quadratic equation:

$$ax^2 + bx + c = 0$$

where $a \neq 0$. The solution(s) to this equation can be found using this formula:

$$x = [-b \pm (b^2 - 4ac)^{1/2}] / 2a$$

The symbol $\pm$ is read "plus-or-minus," and is a way of compacting two mathematical expressions that are additive inverses into a single expression. Written separately, the equations are:

$$x = [-b + (b^2 - 4ac)^{1/2}] / 2a$$
$$x = [-b - (b^2 - 4ac)^{1/2}] / 2a$$

The fractional exponent means the 1/2 power, which is, as you have learned, another way of expressing the positive square root of a quantity.

THE DISCRIMINANT

Consider again the general quadratic equation:

$$ax^2 + bx + c = 0$$

where a, b, and c are real numbers, and $a \neq 0$. Define the discriminant, d, as follows:

$$d = b^2 - 4ac$$

Let the complex-number solutions to the quadratic equation be denoted as follows:

$$x_1 = a_1 + jb_1$$
$$x_2 = a_2 + jb_2$$

Then the following statements hold true:

$$(d > 0) \Rightarrow (b_1 = 0) \ \& \ (b_2 = 0) \ \& \ (a_1 \neq a_2)$$
$$(d = 0) \Rightarrow (b_1 = 0) \ \& \ (b_2 = 0) \ \& \ (a_1 = a_2)$$
$$(d < 0) \Rightarrow (a_1 = a_2) \ \& \ (b_1 = -b_2)$$

The double-shafted arrows are translated as "logically implies." This means that if the expression to the left of the arrow is true, then the expression to the right of the arrow is true. The above three principles are often stated verbally as follows:

- If $d > 0$, then there are two distinct real-number solutions
- If $d = 0$, then there is a single real-number solution
- If $d < 0$, then there are two distinct complex-conjugate solutions

PROBLEM 6-6
Solve the following equation, assuming the values of the coefficients are mathematically exact:

$$-3x^2 - 4x - 2 = 0$$

Express the solution(s) to five significant figures.

SOLUTION 6-6

The coefficients are $a = -3$, $b = -4$, and $c = -2$. Plugging these numbers into the quadratic formula gives us this:

$$x = \{4 \pm [(-4)^2 - (4 \times -3 \times -2)]^{1/2}\} / (2 \times -3)$$
$$= [4 \pm (16 - 24)^{1/2}] / (-6)$$
$$= [4 \pm (-8)^{1/2}] / (-6)$$

We are confronted with the square root of -8. This is equal to the imaginary number $j(8^{1/2})$, which is $j2.8284$ (accurate to five significant figures). The solutions are therefore approximately:

$$x = (4 + j2.8284) / (-6)$$
$$x = (4 - j2.8284) / (-6)$$

These solutions, as complex numbers in their conventional form, are approximately:

$$x = -0.66667 - j0.47140$$
$$x = -0.66667 + j0.47140$$

PROBLEM 6-7

Convert the following quadratic equations into factored form:

$$x^2 - 2x - 15 = 0$$
$$x^2 + 4 = 0$$

SOLUTION 6-7

The first equation turns out to have a "clean" factored equivalent with real-number coefficients:

$$(x + 3)(x - 5) = 0$$

The second equation does not have any real-number solutions. You can tell that something is peculiar about this equation if you subtract 4 from each side:

$$x^2 + 4 = 0$$
$$x^2 = -4$$

No real number can be substituted for x in this equation in order to make it true. However, the equation can be put into complex-number factored form, as follows:

$$(x + j2)(x - j2) = 0$$

PROBLEM 6-8

Put the following factored equations into standard quadratic form:

$$(x + 5)(x - 1) = 0$$
$$x(x + 4) = 0$$

SOLUTION 6-8

Both of these can be converted to standard form by multiplying the factors. In the first case, it goes like this:

$$(x + 5)(x - 1) = 0$$
$$x^2 - x + 5x + (5 \times -1) = 0$$
$$x^2 + 4x - 5 = 0$$

In the second case, it goes like this:

$$x(x + 4) = 0$$
$$x^2 + 4x = 0$$

Higher-Order Equations

As the exponents in single-variable equations get larger, finding the solutions becomes challenging. In the olden days, a lot of insight, guesswork, and tedium was involved in solving such equations. Today, mathematicians and scientists have the help of computers, and when problems are encountered containing equations with variables raised to large powers, they let a computer solve the problem by "brute force." The material here is presented only so you will recognize higher-order equations when you see them.

THE CUBIC

A cubic equation, also called a *one-variable, third-order equation* or a *third-order equation in one variable*, can be written in the following standard form:

$$ax^3 + bx^2 + cx + d = 0$$

where a, b, c, and d are constants, x is the variable, and $a \neq 0$.

If you're lucky, you'll be able to reduce a cubic equation to factored form to find real-number solutions r, s, and t:

$$(x - r)(x - s)(x - t) = 0$$

Don't count on being able to factor a cubic equation. Sometimes it's easy, but most of the time it is difficult. There is a *cubic formula* that can be used in a manner similar to the way in which the quadratic formula is used for quadratic equations, but it's complicated, and will not be presented here.

THE QUARTIC

A *quartic equation*, also called a *one-variable, fourth-order equation* or a *fourth-order equation in one variable*, can be written in the following standard form:

$$ax^4 + bx^3 + cx^2 + dx + e = 0$$

where a, b, c, d, and e are constants, x is the variable, and $a \neq 0$.

Once in a while you will be able to reduce a quartic equation to factored form to find real-number solutions r, s, t, and u:

$$(x - r)(x - s)(x - t)(x - u) = 0$$

As is the case with the cubic, you will be lucky if you can factor a quartic equation into this form and thus find four real-number solutions with ease.

THE QUINTIC

A *quintic equation*, also called a *one-variable, fifth-order equation* or a *fifth-order equation in one variable*, can be written in the following standard form:

$$ax^5 + bx^4 + cx^3 + dx^2 + ex + f = 0$$

where a, b, c, d, e, and f are constants, x is the variable, and $a \neq 0$.

There is a remote possibility that, if you come across a quintic, you'll be able to reduce it to factored form to find real-number solutions r, s, t, u, and v:

$$(x - r)(x - s)(x - t)(x - u)(x - v) = 0$$

As is the case with the cubic and the quartic, you will be lucky if you can factor a quintic equation into this form.

THE *n*th-ORDER EQUATION

A *one-variable, nth-order equation* can be written in the following standard form:

$$a_1 x^n + a_2 x^{n-1} + a_3 x^{n-2} + \ldots + a_{n-2} x^2 + a_{n-1} x + a_n = 0$$

where $a_1, a_2, \ldots a_n$ are constants, x is the variable, and $a_1 \neq 0$. We won't even think about trying to factor an equation like this in general, although specific cases might lend themselves to factorization. Solving *n*th-order equations, where $n > 5$, practically demands the use of a computer.

PROBLEM 6-9

What are the solutions to the following factored equation?

$$(x - j3)(x + 2)(x + j5) = 0$$

SOLUTION 6-9

To solve this, simply identify the three quantities that make any one of the factors equal to 0. These quantities are $j3$ for the first factor, -2 for the second factor, and $-j5$ for the third factor. Mathematicians speak of the set containing the solutions of an equation like this as the *solution set*. If we call that set S in this case, then:

$$S = \{j3, -2, -j5\}$$

PROBLEM 6-10

Write the equation from Problem 6-9 as a cubic equation in standard form. The coefficients will be complex. Be sure these coefficients are expressed in the form $a + jb$, where a and b are real numbers and j is equal to the positive square root of -1.

SOLUTION 6-10

The conversion of a factored equation to standard form is easier than the reverse process, although it can be tedious if there are more than two factors. Let's multiply the second two factors first:

$$(x + 2)(x + j5) = x^2 + j5x + 2x + j10$$

The product of this with the first factor is:

$$(x - j3)(x^2 + j5x + 2x + j10) = x^3 + j5x^2 + 2x^2 + j10x - j3x^2 - j^2 15x - j6x$$
$$- j^2 30$$
$$= x^3 + j5x^2 + 2x^2 + j10x - j3x^2 + 15x - j6x + 30$$
$$= x^3 + (j5 + 2 - j3)x^2 + (j10 + 15 - j6)x + 30$$
$$= x^3 + (2 + j2)x^2 + (15 + j4)x + 30$$

The cubic equation in standard form is therefore:

$$x^3 + (2 + j2)x^2 + (15 + j4)x + 30 = 0$$

Quick Practice

Here are some practice problems that cover the material presented in this chapter. Solutions follow the problems.

PROBLEMS

1. Factor the following quadratic equation:
$$y^2 + 7y + 12 = 0$$

2. State the solution set S for the quadratic equation given in Practice Problem 1.

3. State the following quadratic equation in standard form:
$$x(x - 5) = 0$$

4. What can be said about the solutions to this quadratic equation?
$$2z^2 + 2z + 5 = 0$$

5. State the solution set S for the quadratic equation given in Practice Problem 4. Use the quadratic formula to find the solutions.

SOLUTIONS

1. The process of factoring a quadratic equation involves some intuition. In this case, you might be able to quickly see the factors:

$$(y + 3)(y + 4) = y^2 + 4y + 3y + 12$$
$$= y^2 + 7y + 12$$

so the factored form is:

$$(y + 3)(y + 4) = 0$$

2. It's easy to solve a quadratic or higher-order equation when we see it in factored form. The solutions are those values of the variable that make either of the factors equal to 0. In this case, these values are $y = -3$ and $y = -4$. The solution set is therefore:

$$S = \{-3, -4\}$$

3. Multiply the factors. Here, the first factor is x, and the second factor is $(x - 5)$. Therefore, the side of the equation to the left of the equality symbol is:

$$x(x - 5) = x^2 - 5x$$

and the quadratic equation in standard form is:

$$x^2 - 5x + 0 = 0$$

It is not necessary to write down any addend in the standard form of a quadratic or higher-order equation when the coefficient of that addend is equal to 0. It is perfectly all right, in this case, to write:

$$x^2 - 5x = 0$$

4. Consider the discriminant $d = b^2 - 4ac$, where $a = 2$, $b = 2$, and $c = 5$:

$$d = 2^2 - (4 \times 2 \times 5) = 4 - 40 = -36$$

In this example, $d < 0$, indicating that there are two distinct solutions, and that they are complex conjugates.

5. We already know that $b^2 - 4ac = -36$. The first solution is:

$$z = [-2 + (-36)^{1/2}] / (2 \times 2)$$
$$= (-2 + j6) / 4$$
$$= -2/4 + (j6)/4$$
$$= -1/2 + j(3/2)$$

The second solution is:

$$z = [-2 - (-36)^{1/2}] / (2 \times 2)$$
$$= (-2 - j6) / 4$$
$$= -2/4 - (j6)/4$$
$$= -1/2 - j(3/2)$$

The solution set is therefore:

$$S = \{[-1/2 + j(3/2)], [-1/2 - j(3/2)]\}$$

Quiz

This is an "open book" quiz. You may refer to the text in this chapter. A good score is 8 correct. Answers are in the back of the book.

1. Consider the following equation:

$$z^2 + 25 = 0$$

   How many solutions does this equation have, and what is its, or their, nature?

   (a) There are two distinct real-number solutions.
   (b) There is a single real-number solution.
   (c) There are two complex-conjugate solutions.
   (d) It is impossible to tell without more information.

2. What sort of equation is the following?

$$-21w^2 - 17w + 45 = 6w^2 + 10w + 3$$

   (a) A linear equation.
   (b) A quadratic equation.
   (c) An equation of unknown order.
   (d) An invalid equation.

3. Consider the following equation:

$$x^3 - x = 0$$

   What is this equation, expressed in factored form?

(a) $x(x - 1)(x + 1) = 0$

(b) $x^2 x^3 = 0$

(c) $x + 2x - 3x = 0$

(d) It cannot be expressed in factored form.

4. Consider the following equation:

$$6y + 3 = 8y + 4$$

How many solutions does this equation have, and what is its, or their, nature?

(a) There are two distinct real-number solutions.

(b) There is a single real-number solution.

(c) There are two complex-conjugate solutions.

(d) It is impossible to tell without more information.

5. Consider again the following equation:

$$x^3 - x = 0$$

What is the real-number solution set S for this equation?

(a) $S = \{0\}$

(b) $S = \{0, 1\}$

(c) $S = \{0, -1, 1\}$

(d) $S = \varnothing$, because there are no real-number solutions.

6. Which of the following is *not* a quadratic equation?

(a) $(x + 2)(x - 3) = 5$

(b) $x + 3x - 4 = 3$

(c) $(x + j5)(x - j4) = 8$

(d) None of the above are quadratic equations.

7. Consider the following quadratic equation in standard form:

$$px^2 + qx + r = 0$$

where p, q, and r are real-number constants, and x is the unknown. This equation has a single real-number solution if and only if

(a) $p = q = r$.

(b) $q^2 = 4pr$.

(c) $r^2 < 4pq$.

(d) $r > 4pq$.

8. Consider, yet again, the following equation:

$$x^3 - x = 0$$

It is tempting to suppose that we can divide this equation through on each side by x, getting a simpler equation, as follows:

$$x^3 - x = 0$$
$$(x^3 - x) / x = 0/x$$
$$x^3/x - x/x = 0$$
$$x^2 - 1 = 0$$

Unfortunately, we cannot legitimately do this because

(a) this is not a cubic equation in standard form.
(b) there are no real-number solutions.
(c) it cannot generally be done for higher-order equations.
(d) the solution set of the original equation contains the element 0.

9. In a quadratic equation with two distinct solutions, both of which are real numbers,

(a) the discriminant is a positive real number.
(b) the discriminant is a positive imaginary number.
(c) the discriminant is a negative real number.
(d) the discriminant is a negative imaginary number.

10. Consider the following equation:

$$x^4 = 1$$

Which, if any, of the following numbers is *not* an element of the solution set of this equation?

(a) 1.
(b) −1.
(c) $0 + j1$.
(d) All of three of the above are elements of the solution set of this equation.

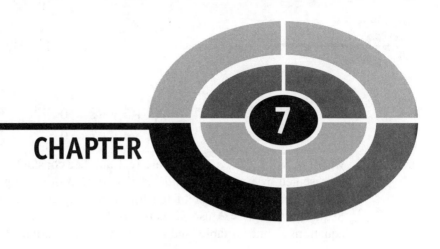

Multivariable Equations

When you want to solve a group of *simultaneous equations*, various schemes can be used. This chapter outlines some of the most common techniques.

2×2 Linear Equations

Let's look at two methods of solving pairs of *simultaneous linear equations*. Either approach can be used for any given pair of linear equations.

2×2 SUBSTITUTION METHOD

Consider the following set of two linear equations in two variables:

$$a_1 x + b_1 y + c_1 = 0$$
$$a_2 x + b_2 y + c_2 = 0$$

where a_1, a_2, b_1, b_2, c_1, and c_2 are real-number constants, and the variables are represented by x and y. The *substitution method* of solving these equations consists in performing either of the following sequences of steps. If $a_1 \neq 0$, use Sequence A. If $a_1 = 0$, use Sequence B. (If $a_1 = 0$ and $a_2 = 0$, then we have two equations in one variable, and the following steps are irrelevant.)

SEQUENCE A

Solve the first equation for x in terms of y:

$$a_1 x + b_1 y + c_1 = 0$$
$$a_1 x = -b_1 y - c_1$$
$$x = (-b_1 y - c_1) / a_1$$

Next, substitute the above-derived solution for x in the second equation:

$$a_2[(-b_1 y - c_1) / a_1] + b_2 y + c_2 = 0$$

Solve this single-variable linear equation for y, using the "operational rules" from the previous chapter. Assuming a solution exists, it can be substituted for y in either of the original equations, deriving a single-variable equation in terms of x. Once this has been done, solve that equation for x, using the "operational rules" outlined in the previous chapter.

SEQUENCE B

Because $a_1 = 0$, the first equation has only one variable, and is in the following form:

$$b_1 y + c_1 = 0$$

Solve this equation for y:

$$b_1 y = -c_1$$
$$y = -c_1 / b_1$$

This can be substituted for y in the second equation, obtaining:

$$a_2x + b_2(-c_1/b_1) + c_2 = 0$$
$$a_2x - b_2(c_1/b_1) + c_2 = 0$$
$$a_2x = b_2(c_1/b_1) - c_2$$
$$x = [b_2(c_1/b_1) - c_2] / a_2$$

2×2 ADDITION METHOD

Consider the following set of two linear equations in two variables:

$$a_1x + b_1y + c_1 = 0$$
$$a_2x + b_2y + c_2 = 0$$

where a_1, a_2, b_1, b_2, c_1, and c_2 are real-number constants, and the variables are represented by x and y. The *addition method* of solving these equations consists in performing two separate and independent steps:

- Multiply one or both equations through by a constant, if necessary, to cancel out the coefficients of x in the sum of the two equations, and then solve the sum of the two equations for y.
- Multiply one or both equations through by a constant, if necessary, to cancel out the coefficients of y in the sum of the two equations, and then solve the sum of the two equations for x.

To solve for y, begin by multiplying the first equation through by $-a_2$, and the second equation through by a_1, and then add the two resulting equations:

$$(-a_2a_1x - a_2b_1y - a_2c_1 = 0)$$
$$+ (a_1a_2x + a_1b_2y + a_1c_2 = 0)$$
$$\overline{(a_1b_2 - a_2b_1)y + a_1c_2 - a_2c_1 = 0}$$

Then, add a_2c_1 to each side, obtaining:

$$(a_1b_2 - a_2b_1)y + a_1c_2 = a_2c_1$$

Next, subtract a_1c_2 from each side, obtaining:

$$(a_1b_2 - a_2b_1)y = a_2c_1 - a_1c_2$$

Finally, divide through by $a_1b_2 - a_2b_1$, obtaining:

$$y = (a_2c_1 - a_1c_2) / (a_1b_2 - a_2b_1)$$

For this to be valid, the denominator must be nonzero. This requires that $a_1b_2 \neq a_2b_1$. If it turns out that $a_1b_2 = a_2b_1$, then there are not two distinct solutions to the set of equations.

The process of solving for x is similar. Consider again the original equations:

$$a_1x + b_1y + c_1 = 0$$
$$a_2x + b_2y + c_2 = 0$$

Multiply the first equation through by $-b_2$, and the second equation through by b_1, and then add the two resulting equations:

$$(-a_1b_2x - b_1b_2y - b_2c_1 = 0)$$
$$+ (a_2b_1x + b_1b_2y + b_1c_2 = 0)$$
$$\overline{(a_2b_1 - a_1b_2)x + b_1c_2 - b_2c_1 = 0}$$

Then, add b_2c_1 to each side, obtaining:

$$(a_2b_1 - a_1b_2)x + b_1c_2 = b_2c_1$$

Next, subtract b_1c_2 from each side, obtaining:

$$(a_2b_1 - a_1b_2)x = b_2c_1 - b_1c_2$$

Finally, divide through by $a_2b_1 - a_1b_2$, obtaining:

$$x = (b_2c_1 - b_1c_2) / (a_2b_1 - a_1b_2)$$

For this to be valid, the denominator must be nonzero. This requires that $a_1b_2 \neq a_2b_1$. If it turns out that $a_1b_2 = a_2b_1$, then there are not two distinct solutions to the set of equations.

PROBLEM 7-1

Solve the following pair of linear equations (if there is a unique solution) using the substitution method. If no solution exists, or if there are infinitely many solutions, then say so.

$$3x - 6y + 9 = 0$$
$$-10x - 5y + 15 = 0$$

SOLUTION 7-1

Let's begin by solving for x in terms of y in the first equation. Proceed as follows:

$$3x - 6y + 9 = 0$$
$$3x - 6y = -9$$
$$3x = 6y - 9$$
$$x = 2y - 3$$

Next, substitute the above-derived solution for x in the second equation, obtaining:

$$-10x - 5y + 15 = 0$$
$$-10(2y - 3) - 5y + 15 = 0$$
$$-20y + 30 - 5y + 15 = 0$$
$$-25y + 45 = 0$$
$$-25y = -45$$
$$25y = 45$$
$$y = 45/25 = 9/5$$

Now that we have the solution for y, we can plug it into either of the original equations and solve for x. Let's use the second equation. Then:

$$-10x - 5y + 15 = 0$$
$$-10x - 5(9/5) + 15 = 0$$
$$-10x - 9 + 15 = 0$$
$$-10x + 6 = 0$$
$$-10x = -6$$
$$10x = 6$$
$$x = 6/10 = 3/5$$

PROBLEM 7-2

Solve the following pair of linear equations (if there is a unique solution) using the addition method. If no solution exists, or if there are infinitely many solutions, then say so.

$$-4x + y - 8 = 0$$
$$-8x + 2y - 14 = 0$$

SOLUTION 7-2

Multiply one of the equations through by a constant, such that one of the variables is canceled out when the two equations are added. Let's attempt to cancel out x, getting an equation in y only. If we multiply the first equation through by -2 and then add it to the second one, we get:

$$(8x - 2y + 16 = 0)$$
$$+ (-8x + 2y - 14 = 0)$$
$$\overline{\qquad 2 = 0 \qquad}$$

We've derived a contradiction. Is there some mistake? Let's try to cancel out y instead, multiplying the second equation through by $-1/2$ and then adding the two equations. In this case we get:

$$(-4x + y - 8 = 0)$$
$$+ (4x - y + 7 = 0)$$
$$\overline{\qquad -1 = 0 \qquad}$$

Again, a contradiction! There is no mistake here. Whenever you use the addition method in an attempt to solve a pair of linear equations and get an equation to the effect that one number equals another, it means that the pair of equations has no solution. Such a pair of equations in two variables is said to be *inconsistent*.

Occasionally, when applying the addition method to a pair of linear equations, you'll get another sort of unexpected result: a trivial statement that a number equals itself, such as $0 = 0$. When this happens, it means that the equations are actually different expressions for the same two-variable equation. In that case, the "pair" of equations has infinitely many solutions.

3×3 Linear Equations

A set of three linear equations in three variables presents a more involved (and tedious) problem. The same two general methods, addition and substitution, can be used, but in combination. The following two problems and solutions provide specific examples.

PROBLEM 7-3
Solve the following set of 3×3 linear equations. If no solution exists, or if there are infinitely many solutions, then say so.

$$2x + 5y - z = 0$$
$$-3x - 4y + 2z = 0$$
$$x + y + z = 0$$

SOLUTION 7-3

First, let's cancel out x, getting an equation in y and z. The coefficients of x in the above equations are 2, −3, and 1, respectively. When these three coefficients are added, the result is 0. This is convenient, because we can add the three equations together just as they are in order to cancel x. Proceed:

$$(2x + 5y - z = 0)$$
$$+ (-3x - 4y + 2z = 0)$$
$$+ (x + y + z = 0)$$
$$\overline{}$$
$$2y + 2z = 0$$

This result can be manipulated to solve for z in terms of y:

$$2y + 2z = 0$$
$$y + z = 0$$
$$z = -y$$

Now we can reduce the original 3×3 set of equations (in x, y, and z) into a 2×2 set of equations (in x and y). Let's substitute $-y$ for z in the first two of the original three equations. This gives us:

$$2x + 5y - (-y) = 0$$
$$-3x - 4y - 2y = 0$$

which simplifies to this pair of equations:

$$2x + 6y = 0$$
$$-3x - 6y = 0$$

These can be directly added, obtaining the single-variable equation:

$$-x = 0$$

Multiplying this through by −1 tells us that $x = 0$. That's one variable down, and two to go! Let's substitute 0 for x in the first of the above pair of equations in x and y. This gives us:

$$2x + 6y = 0$$
$$(2 \times 0) + 6y = 0$$
$$6y = 0$$
$$y = 0$$

Now we can substitute $x = 0$ and $y = 0$ into any of the original three equations. Let's use the last one:

$$x + y + z = 0$$
$$0 + 0 + z = 0$$
$$z = 0$$

The solution to this set of 3×3 linear equations, expressed as an ordered triple, is $(x,y,z) = (0,0,0)$.

  Did this solution seem "too easy to be true"? In this case, the three equations were selected because there are no constants added or subtracted to any of the three equations. That always makes a 3×3 set of linear equations comparatively easy to solve. Now let's try a more difficult problem.

PROBLEM 7-4

Solve the following set of 3×3 linear equations. If no solution exists, or if there are infinitely many solutions, then say so.

$$2x + 5y - z + 2 = 0$$
$$-3x - 4y + 2z - 1 = 0$$
$$x + y + z + 5 = 0$$

SOLUTION 7-4

First, let's cancel out x, getting an equation in y and z. The coefficients of x in the above equations are 2, −3, and 1, respectively. When these three coefficients are added, the result is 0, just as it was in the previous problem. Thus:

$$(2x + 5y - z + 2 = 0)$$
$$+ (-3x - 4y + 2z - 1 = 0)$$
$$+ \underline{(x + y + z + 5 = 0)}$$
$$2y + 2z + 6 = 0$$

This result can be manipulated to solve for z in terms of y:

$$2y + 2z + 6 = 0$$
$$2y + 2z = -6$$
$$y + z = -3$$
$$z = -y - 3$$

Now we can reduce the original 3×3 set of equations (in x, y, and z) into a 2×2 set of equations (in x and y). Let's substitute the quantity $(-y - 3)$ for z in the first two of the original three equations. This gives us:

$$2x + 5y - (-y - 3) + 2 = 0$$
$$-3x - 4y + 2(-y - 3) - 1 = 0$$

which simplifies to this pair of equations:

$$2x + 6y + 5 = 0$$
$$-3x - 6y - 7 = 0$$

These can be directly added, obtaining the single-variable equation:

$$-x - 2 = 0$$
$$-x = 2$$

Multiplying this through by -1 tells us that $x = -2$. Let's substitute -2 for x in the first of the above pair of equations in x and y. This gives us:

$$2x + 6y + 5 = 0$$
$$(2 \times -2) + 6y + 5 = 0$$
$$-4 + 6y + 5 = 0$$
$$6y + 1 = 0$$
$$6y = -1$$
$$y = -1/6$$

Now we can substitute $x = -2$ and $y = -1/6$ into any of the original three equations. Let's use the last one:

$$x + y + z + 5 = 0$$
$$-2 - 1/6 + z + 5 = 0$$
$$z - 2 - 1/6 + 5 = 0$$
$$z + 17/6 = 0$$
$$z = -17/6$$

The solution to this set of 3×3 linear equations, expressed as an ordered triple, is:

$$(x,y,z) = (-2,-1/6, -17/6)$$

CHECK IT OUT!

After solving a set of equations such as that in Problem 7-4, you should always "plug in the solutions" to each of the original equations, and be sure they check out correctly. With all the adding, subtracting, multiplying, and dividing that goes on in these sorts of calculations, along with changes of sign and double-negatives, it's easy to make mistakes. As an exercise, check out the above solutions. Plug $x = -2$, $y = -1/6$, and $z = -17/6$ into each of the original three equations, and see if the results are valid.

2×2 General Equations

When one or both of the equations in a 2×2 set are nonlinear, the substitution method generally works best. Two examples follow.

EXAMPLE A

Consider the following two equations:

$$y = x^2 + 2x + 1$$
$$y = -x + 1$$

The first equation is quadratic, and the second equation is linear. Either equation can be directly substituted into the other to solve for x. Substituting the second equation into the first yields this result.

$$-x + 1 = x^2 + 2x + 1$$

This equation can be put into standard quadratic form as follows:

$$-x + 1 = x^2 + 2x + 1$$
$$-x = x^2 + 2x$$
$$0 = x^2 + 3x$$
$$x^2 + 3x = 0$$

We can use the quadratic formula to solve this. Call the coefficients in the previous quadratic equation $a = 1$, $b = 3$, and $c = 0$. Call the solutions x_1 and x_2. Then:

$$x = [-3 \pm (3^2 - 4 \times 1 \times 0)^{1/2}] / (2 \times 1)$$
$$x = [-3 \pm (9 - 0)^{1/2}] / 2$$
$$x = [-3 \pm 9^{1/2}] / 2$$
$$x = (-3 \pm 3) / 2$$
$$x_1 = -6/2 \text{ and } x_2 = 0/2$$
$$x_1 = -3 \text{ and } x_2 = 0$$

These values can be substituted into the original linear equation to obtain the solutions for y, which we call y_1 and y_2 to distinguish them from each other:

$$y_1 = -(-3) + 1 \text{ and } y_2 = -0 + 1$$
$$y_1 = 3 + 1 \text{ and } y_2 = 0 + 1$$
$$y_1 = 4 \text{ and } y_2 = 1$$

The solutions are therefore:

$$(x_1, y_1) = (-3, 4)$$
$$(x_2, y_2) = (0, 1)$$

EXAMPLE B

Consider the following two equations:

$$y = -2x^2 + 4x - 5$$
$$y = -2x - 5$$

The first equation is quadratic, and the second equation is linear. Either equation can be directly substituted into the other to solve for x. Substituting the second equation into the first yields this result:

$$-2x - 5 = -2x^2 + 4x - 5$$

This equation can be put into standard quadratic form as follows:

$$-2x - 5 = -2x^2 + 4x - 5$$
$$-2x = -2x^2 + 4x$$
$$0 = -2x^2 + 6x$$
$$-2x^2 + 6x = 0$$

Use the quadratic formula, and let $a = -2$, $b = 6$, and $c = 0$:

$$x = [-6 \pm (6^2 - 4 \times -2 \times 0)^{1/2}] / (2 \times -2)$$
$$x = [-6 \pm (36 - 0)^{1/2}] / -4$$
$$x = [-6 \pm 36^{1/2}] / -4$$
$$x = (-6 \pm 6) / -4$$
$$x_1 = 0/-4 \text{ and } x_2 = -12/-4$$
$$x_1 = 0 \text{ and } x_2 = 3$$

These values can be substituted into the original linear equation to obtain the y-values:

$$y_1 = -2 \times 0 - 5 \text{ and } y_2 = -2 \times 3 - 5$$
$$y_1 = -5 \text{ and } y_2 = -11$$

The solutions are therefore:

$$(x_1, y_1) = (0, -5)$$
$$(x_2, y_2) = (3, -11)$$

Graphic Solution of Pairs of Equations

The solutions of pairs of equations in two variables can be approximated by graphing the equations. In a graph, the solutions to a set of equations appear as intersection points between the graphs of the equations. If two graphs do not intersect, then the pair of equations they represent has no solution. If the two graphs coincide, then the pair of equations they represent has infinitely many solutions.

EXAMPLE A

Refer to Example A on page 152. Consider this pair of equations:

$$y = x^2 + 2x + 1$$
$$y = -x + 1$$

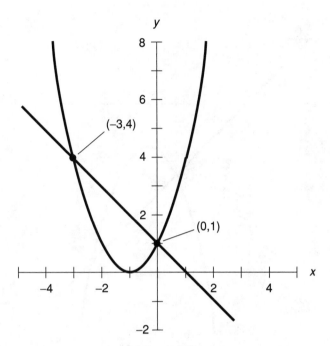

Fig. 7-1. A graph showing the solutions of $y = x^2 + 2x + 1$ and $y = -x + 1$.

These equations are graphed in Fig. 7-1. The graph of the quadratic equation is a curve known as a *parabola*, and the graph of the linear equation is a straight line. The line crosses the parabola at two points, indicating that there are two solutions of this pair of equations. The coordinates of the points, corresponding to the solutions, are:

$$(x_1, y_1) = (-3, 4)$$
$$(x_2, y_2) = (0, 1)$$

EXAMPLE B

Refer to Example B on page 153. Consider this pair of equations:

$$y = -2x^2 + 4x - 5$$
$$y = -2x - 5$$

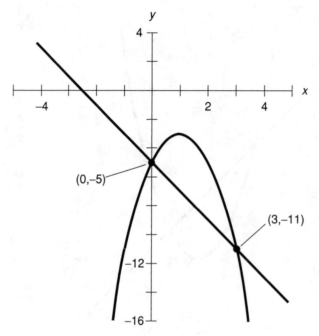

Fig. 7-2. A graph showing the solutions of $y = -2x^2 + 4x - 5$ and $y = -2x - 5$.

These equations are graphed in Fig. 7-2. The line crosses the parabola at two points, indicating that there are two solutions of this set of simultaneous equations. The coordinates of the points, corresponding to the solutions, are:

$$(x_1, y_1) = (0, -5)$$
$$(x_2, y_2) = (3, -11)$$

PROBLEM 7-5

Convert the pair of linear equations from Problem 7-2 into slope-intercept form and graph them in the Cartesian plane. Then explain how this graph portrays the fact that this pair of simultaneous linear equations has no common solution. Again, this pair of equations is:

$$-4x + y - 8 = 0$$
$$-8x + 2y - 14 = 0$$

SOLUTION 7-5
Using the rules of manipulation for linear equations, the first equation can be changed into slope-intercept form like this:

$$-4x + y - 8 = 0$$
$$y - 8 = 4x$$
$$y = 4x + 8$$

This means that the slope is equal to 4, and the y intercept is equal to 8. This line, with a slope of 4, increases by 4 units along the y axis for every unit increase along the x axis. Because the y intercept is 8, this line crosses the y axis at the point (0,8). Its graph is shown in Fig. 7-3 by the solid line.

The second equation is converted to slope-intercept form as follows:

$$-8x + 2y - 14 = 0$$
$$2y - 14 = 8x$$
$$2y = 8x + 14$$
$$y = 4x + 7$$

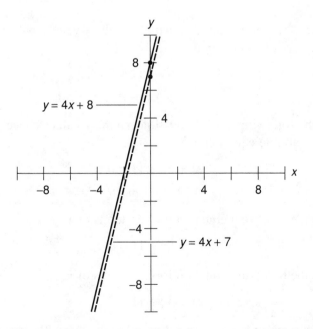

Fig. 7-3. Illustration for Problem 7-5.

This means that the slope is equal to 4, and the y intercept is equal to 7. This line, with a slope of 4, increases by 4 units along the y axis for every unit increase along the x axis (the same situation as with the other line). Because the y intercept is 7, this line crosses the y axis at the point (0,7). Its graph is shown in Fig. 7-3 by the dashed line.

Note that the slopes of the two lines are the same, but their y intercepts differ. Any two lines that have the same slope but different y intercepts are parallel, so they never intersect in the Cartesian plane. The fact that the graphs of the equations have no point in common portrays the fact that the pair of equations has no solution.

Quick Practice

Here are some practice problems that cover the material presented in this chapter. Solutions follow the problems.

PROBLEMS

1. Add the following pair of equations so the x values cancel out, giving a single-variable equation in y:

$$2x + 5y + 11 = 0$$
$$5x - y + 4 = 0$$

2. Add the following pair of equations so the y values cancel out, giving a single-variable equation in x:

$$2x + 5y + 11 = 0$$
$$5x - y + 4 = 0$$

3. Solve the following equation for y in terms of x:

$$2x + 5y + 11 = 0$$

4. Solve the following equation for x in terms of y:

$$2x + 5y + 11 = 0$$

5. Using the result in Practice Solution 1, solve for y. Reduce the answer to the simplest possible form.

SOLUTIONS

1. Multiply the top equation through by −5 and the bottom equation through by 2, and then add the resulting equations. This gives an equation in y, as follows:

$$(-10x - 25y - 55 = 0)$$
$$+ (10x - 2y + 8 = 0)$$
$$\overline{-27y - 47 = 0}$$

2. Multiply the bottom equation through by 5, and then add the resulting equations. This gives an equation in x, as follows:

$$(2x + 5y + 11 = 0)$$
$$+ (25x - 5y + 20 = 0)$$
$$\overline{27x + 31 = 0}$$

3. This can be done algebraically, using the principles of calculation in Chapter 2 and the operational rules in Chapter 6:

$$2x + 5y + 11 = 0$$
$$2x + 5y = -11$$
$$5y = -11 - 2x$$
$$y = (-11 - 2x) / 5$$
$$y = -11/5 - (2/5)x$$

4. This can be done algebraically, using the principles of calculation in Chapter 2 and the operational rules in Chapter 6:

$$2x + 5y + 11 = 0$$
$$2x + 5y = -11$$
$$2x = -11 - 5y$$
$$x = (-11 - 5y) / 2$$
$$x = -11/2 - (5/2)y$$

5. Manipulate the results of Practice Solution 1, as follows:

$$-27y - 47 = 0$$
$$-27y = 47$$
$$y = 47 / (-27)$$
$$y = -47/27$$

Quiz

This is an "open book" quiz. You may refer to the text in this chapter. A good score is 8 correct. Answers are in the back of the book.

1. Figure 7-4 portrays the graphs of three different equations in two variables, x and y. The graphs are called P, Q, and R. We don't know exactly what the equations are, but we can nevertheless tell from the graphs that

   (a) there is no real-number solution to all three of these equations considered simultaneously.
   (b) there is exactly one real-number solution to all three of these equations considered simultaneously.
   (c) there are two distinct real-number solutions to all three of these equations considered simultaneously.
   (d) there are three real-number solutions to all three of these equations considered simultaneously.

2. In Fig. 7-4, how many real-number solutions exist between the equations represented by curve P and line Q?

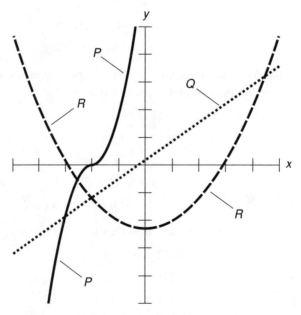

Fig. 7-4. Illustration for Quiz Questions 1 through 4.

(a) None

(b) One

(c) Two

(d) Three

3. In Fig. 7-4, how many real-number solutions exist between the equations represented by curve P and curve R?

(a) None

(b) One

(c) Two

(d) Three

4. In Fig. 7-4, how many real-number solutions exist between the equations represented by line Q and curve R?

(a) None

(b) One

(c) Two

(d) Three

5. Suppose you encounter a pair of linear equations in two variables. You reduce them to slope-intercept form and find that their slopes are the same, but their x intercepts (the points where the graphs intersect the x axis) are different. From this, you can conclude that

(a) there is no real-number solution to this pair of equations.

(b) there is exactly one real-number solution to this pair of equations.

(c) there are exactly two distinct real-number solutions to this pair of equations.

(d) there are infinitely many real-number solutions to this pair of equations.

6. Suppose you encounter a pair of linear equations in two variables. You reduce them to slope-intercept form and find that their slopes are different, and their x intercepts are also different. From this, you can conclude that

(a) there is no real-number solution to this pair of equations.

(b) there is exactly one real-number solution to this pair of equations.

(c) there are exactly two distinct real-number solutions to this pair of equations.

(d) there are infinitely many real-number solutions to this pair of equations.

7. Consider the following two linear equations in variables x and y, where a, b, and c are constants, and $a \neq b$:

$$y = ax + c$$
$$y = bx + c$$

From this information, what can we conclude about these two equations?

(a) Their graphs have the same slope and the same y intercept.
(b) Their graphs have the same slope but different y intercepts.
(c) In the real-number solution to these equations, $x = c$.
(d) In the real-number solution to these equations, $y = c$.

8. Consider the following two linear equations in variables r and s, where f, g, and h are constants, and $g \neq h$:

$$s = fr + g$$
$$s = fr + h$$

From this information, what can we conclude about these two equations?

(a) Their graphs have the same slope.
(b) Their graphs have the same y intercept.
(c) There exists exactly one real-number solution to this pair of equations.
(d) None of the above

9. Imagine two equations in two variables. When graphed in the Cartesian xy-plane, one of the graphs is a parabola that opens upward, and attains a minimum y value (it "bottoms out") at the point $(0,3)$. The other parabola opens downward, and attains a maximum y value (it "peaks") at the point $(0,-3)$. From this information, you can be certain that

(a) their graphs have the same slope.
(b) there is no real-number solution to this pair of equations.
(c) there is exactly one real-number solution to this pair of equations.
(d) there are two distinct real-number solutions to this pair of equations.

10. Imagine two equations in two variables. When graphed in the Cartesian xy-plane, one of the graphs is a circle centered at the origin $(0,0)$. The other graph is a straight line passing through the origin. From this information, you can be certain that

(a) their graphs have the same slope.
(b) there is no real-number solution to this pair of equations.
(c) there is exactly one real-number solution to this pair of equations.
(d) there are two distinct real-number solutions to this pair of equations.

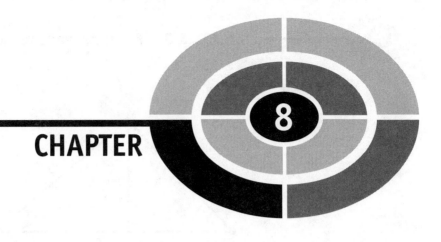

Perimeter and Area in Two Dimensions

In this chapter, we'll review formulas for calculating the perimeters and interior areas of common geometric figures on flat (Euclidean) surfaces.

Triangles

A *triangle* is a geometric figure defined by three distinct points that do not all lie along the same line. Each point is a *vertex* of the triangle. The line segments connecting pairs of points compose the *sides* of the triangle. The angles between adjacent sides, expressed inside the triangle, are the *interior angles*. The sum of the measures of the interior angles of any triangle in a geometric plane is equal to 180° (π rad).

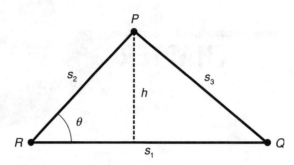

Fig. 8-1. Dimensions of a triangle.

PERIMETER OF TRIANGLE

Suppose T is a triangle defined by points P, Q, and R, and having sides of lengths s_1, s_2, and s_3 as shown in Fig. 8-1. Let s_1 be the base length, and let h be the height. The perimeter, B, of the triangle is given by the following formula:

$$B = s_1 + s_2 + s_3$$

INTERIOR AREA OF TRIANGLE

Let T be a triangle as defined above and in Fig. 8-1. Let θ be the measure of the interior angle between the sides having lengths s_1 and s_2. The interior area, A, of the triangle is given by either of the following formulas:

$$A = s_1 h/2$$
$$A = (s_1 s_2 \sin \theta)/2$$

where the abbreviation *sin* represents the *sine function*. (The *trigonometric functions* are discussed in more detail in Chapter 11.)

THEOREM OF PYTHAGORAS

Suppose T is a *right triangle*, which is a triangle in which one of the interior angles is a *right angle*. Suppose T is defined by three points P, Q, and R, and the sides of T have lengths s_1, s_2, and s_3. Let s_3 be the length of the *hypotenuse* (the side opposite the right angle, and always the longest side in a right triangle). The following equation applies:

$$s_3 = (s_1{}^2 + s_2{}^2)^{1/2}$$

The converse of this is also true: If T is a triangle whose sides have lengths s_1, s_2, and s_3 such that the above equation holds, then T is a right triangle.

PROBLEM 8-1

Suppose you are 160 cm tall. You stand in the sun, and your shadow falls on a horizontal, flat surface. Your shadow measures 220 cm in length. What is the area, in square meters (m^2), of the triangle defined by your feet, the top of your head, and the end of your shadow?

SOLUTION 8-1

Use the formula for the interior area of a triangle, letting the length of your shadow be s_1 and your height be s_2. Note that s_2 is equal to the height of the triangle in this case, so $s_2 = h$ in the first formula for the interior area of a triangle. Convert the values given into meters: $s_1 = 220$ cm $= 2.20$ m, and $h = 160$ cm $= 1.60$ m. Then:

$$A = s_1 h / 2$$
$$= 2.20 \times 1.60 / 2$$
$$= 2.20 \times 0.800$$
$$= 1.76 \ m^2$$

PROBLEM 8-2

What is the perimeter of the triangle in the above scenario, in centimeters?

SOLUTION 8-2

Note that the defined triangle is a right triangle (assuming you stand vertically), because the surface on which you stand is flat and horizontal. The angle between the side of length s_1 (the length of your shadow) and the side of length s_2 (your height) is a right angle. First, determine the distance from the top of your head to the end of your shadow. This is s_3 in the formula for the Theorem of Pythagoras. Using the values given in centimeters, proceed as follows:

$$s_3 = (s_1{}^2 + s_2{}^2)^{1/2}$$
$$= (220^2 + 160^2)^{1/2}$$
$$= (48,400 + 25,600)^{1/2}$$
$$= 74,000^{1/2}$$
$$= 272 \ cm$$

Next, add this to the lengths of the other two sides to get the perimeter:

$$B = 272 + 220 + 160$$
$$= 652 \text{ cm}$$

Quadrilaterals

A *quadrilateral* is a geometric figure defined by four distinct points that all lie in the same plane, but no three of which lie along a single straight line. Each point is a *vertex* of the quadrilateral. The line segments connecting adjacent pairs of points compose the *sides* of the quadrilateral. The angles between adjacent sides, defined inside the figure, are called the *interior angles*. The sum of the measures of the interior angles of any quadrilateral in a plane is equal to 360° (2π rad).

PERIMETER OF PARALLELOGRAM

A *parallelogram* is a quadrilateral such that both pairs of opposite sides are parallel. Whenever both pairs of opposite sides in a quadrilateral are parallel, those pairs have the same length, and both pairs of opposite angles have equal measure. Suppose V is a parallelogram defined by points P, Q, R, and S, and having sides of lengths s_1 and s_2 as shown in Fig. 8-2. Let s_1 be the base length, and let h be the height. The perimeter, B, of the parallelogram is given by the following formula:

$$B = 2s_1 + 2s_2$$

INTERIOR AREA OF PARALLELOGRAM

Let V be a parallelogram as defined above and in Fig. 8-2. Let θ be the measure of any one of the interior angles. The interior area, A, is given by either of these formulas:

$$A = s_1 h$$
$$A = s_1 s_2 \sin \theta$$

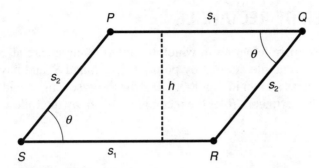

Fig. 8-2. Dimensions of a parallelogram. If $s_1 = s_2$, the figure is a rhombus.

PERIMETER OF RHOMBUS

A *rhombus* is a parallelogram in which all four sides have the same length. Suppose V is a rhombus defined by points P, Q, R, and S, and having sides of length s. The rhombus is a special case of the parallelogram (Fig. 8-2) in which the following holds:

$$s_1 = s_2 = s$$

The perimeter, B, of the rhombus is given by the following formula:

$$B = 4s$$

INTERIOR AREA OF RHOMBUS

Let V be a rhombus as defined above and in Fig. 8-2. Let d_1 and d_2 be the lengths of the diagonals of the rhombus. The interior area, A, is given by any of these formulas:

$$A = sh$$
$$A = s^2 \sin \theta$$
$$A = d_1 d_2 / 2$$

PERIMETER OF RECTANGLE

A *rectangle* is a parallelogram in which the interior angles are all right angles. Suppose V is a rectangle defined by points P, Q, R, and S, and having sides of lengths s_1 and s_2 as shown in Fig. 8-3. Let s_1 be the base length, and let s_2 be the height. Then the perimeter, B, of the rectangle is given by the following formula:

$$B = 2s_1 + 2s_2$$

INTERIOR AREA OF RECTANGLE

Let V be a rectangle as defined above and in Fig. 8-3. The interior area, A, is given by:

$$A = s_1 s_2$$

PERIMETER OF SQUARE

A *square* is a rectangle in which all four sides have the same length, or a rhombus in which the interior angles are all right angles. Suppose V is a square defined by points P, Q, R, and S, and having sides of length s. This is a special case of the rectangle (Fig. 8-3) in which the following holds:

$$s_1 = s_2 = s$$

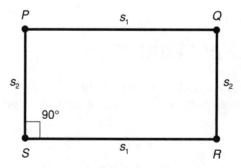

Fig. 8-3. Dimensions of a rectangle. If $s_1 = s_2$, the figure is a square.

The perimeter, B, of the square is given by the following formula:

$$B = 4s$$

INTERIOR AREA OF SQUARE

Let V be a square as defined above and in Fig. 8-3. Let d be the length of the diagonal of the square. The interior area, A, is given by either of these formulas:

$$A = s^2$$
$$A = d^2/2$$

PERIMETER OF TRAPEZOID

A *trapezoid* is a quadrilateral in which one pair of opposite sides is parallel, but there are no other constraints. Suppose V is a trapezoid defined by points P, Q, R, and S, and having sides of lengths s_1, s_2, s_3, and s_4 as shown in Fig. 8-4. Let s_1 be the base length, let h be the height, let θ be the measure of the angle between the sides having length s_1 and s_2, and let ϕ be the measure of the angle between the sides having length s_1 and s_4. Suppose the sides having lengths s_1 and s_3 (line segments PQ and RS) are parallel. The perimeter, B, of the trapezoid is given by either of the following formulas:

$$B = s_1 + s_2 + s_3 + s_4$$
$$B = s_1 + s_3 + h \csc \theta + h \csc \phi$$

where the abbreviation *csc* represents the *cosecant function*. The cosecant of an angle is equal to the reciprocal of the sine (1 divided by the sine) of that angle. Thus, the second formula above can be rewritten in "calculator-friendly" form like this:

$$B = s_1 + s_3 + h/(\sin \theta) + h/(\sin \phi)$$

INTERIOR AREA OF TRAPEZOID

Let V be a trapezoid as defined above and in Fig. 8-4. The interior area, A, is given by the following formula:

$$A = (s_1 h + s_3 h)/2$$

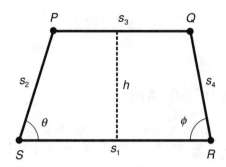

Fig. 8-4. Dimensions of a trapezoid.

PROBLEM 8-3

Suppose you want to carpet a room that is shaped like a parallelogram. The long walls in the room are each 14.0 feet (ft) in length as measured along the floor, while the short walls are each 10.0 ft in length. The acute angles between the short walls and the long walls measure 70.0°. How much carpet will you need?

SOLUTION 8-3

Refer to Fig. 8-2. Let s_1 be the length of the long wall, let s_2 be the length of the short wall, and θ be the measure of the acute angle between the walls. Then $s_1 = 14.0$ and $s_2 = 10.0$. We are given $\theta = 70.0°$. We can use the following formula to find the area of the floor in square feet (ft^2), which is the amount of carpet we will need:

$$A = s_1 s_2 \sin \theta$$
$$= 14.0 \times 10.0 \times \sin 70.0°$$
$$= 140 \times 0.9397$$
$$= 132 \text{ ft}^2$$

PROBLEM 8-4

Suppose you want to carpet a room that is shaped like a rectangle. The long walls in the room are each 14.0 ft in length (as measured along the floor), while the short walls are each 10.0 ft in length. How much carpet will you need?

SOLUTION 8-4
Multiply the length of the long wall by the length of the short wall:

$$A = s_1 s_2$$
$$= 14.0 \times 10.0$$
$$= 140 \text{ ft}^2$$

Regular Polygons

A *regular polygon* is a multisided figure defined by n points, where n is an integer greater or equal to 3, and such that all the points lie in a single plane, all the line segments connecting adjacent pairs of points have the same length, and all the interior angles of the figure have equal measure.

PERIMETER OF REGULAR POLYGON

Suppose V is a regular polygon having n sides of length s, and whose vertices are $P_1, P_2, P_3, \ldots, P_n$ as shown in Fig. 8-5, and whose interior angles all have equal measures. The perimeter, B, of the polygon is given by the following formula:

$$B = ns$$

INTERIOR AREA OF REGULAR POLYGON

Let V be a regular polygon as defined above and in Fig. 8-5. The interior area, A, of the polygon is given by the following formula:

$$A = (ns^2/4) \cot (180/n)°$$

where the abbreviation *cot* represents the *cotangent function*. The cotangent of an angle is equal to the reciprocal of the *tangent* (1 divided by the tangent) of that angle. Thus, the second formula above can be rewritten in "calculator-friendly" form like this:

$$A = (ns^2/4) / [\tan (180/n)°]$$

where the abbreviation *tan* represents the *tangent function*.

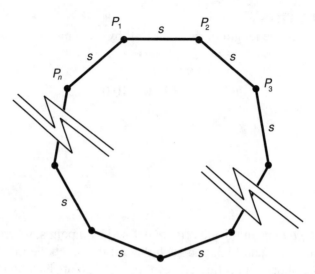

Fig. 8-5. Dimensions of a regular polygon.

PROBLEM 8-5
What is the interior area of a regular octagon whose sides each measure 1.00 m in length? (An octagon is an eight-sided polygon.)

SOLUTION 8-5
Use the formula for the area of a regular polygon, where $n = 8$ (an exact value) and $s = 1.00$ (to three significant figures):

$$A = (ns^2/4) \cot (180/n)°$$
$$= (8 \times 1.00^2/4) \cot (180/8)°$$
$$= 2.00 \cot 22.5°$$
$$= 2.00 \times 2.41421$$
$$= 4.83 \text{ m}^2$$

Circles and Ellipses

Circles and ellipses, and sections of these figures, differ from plane polygons in the sense that some or all of their "sides" are curves. The constant π generally appears in formulas used to evaluate the dimensions of these objects.

CIRCUMFERENCE OF CIRCLE

Suppose C is a circle having radius r as shown in Fig. 8-6. The circumference, B, of the circle is given by the following formula:

$$B = 2\pi r$$

where π is the ratio of the circumference of a circle to its diameter, a constant whose value can be taken as 3.14159 for the purposes of most calculations. Alternatively, if d is the diameter of the circle, the circumference B can be found by this formula:

$$B = \pi d$$

INTERIOR AREA OF CIRCLE

Let C be a circle as defined above, and as shown in Fig. 8-6. The interior area, A, of the circle is given by the following formula:

$$A = \pi r^2$$

Alternatively, if d is the diameter of the circle (always exactly equal to twice the radius), the interior area, A, of the circle can be found by means of this formula:

$$A = \pi d^2/4$$

Fig. 8-6. Dimensions of a circle.

INTERIOR AREA OF ELLIPSE

In an ellipse, there are two important specifications called the *semi-axes*. The *major semi-axis* is half of a line segment passing through the center and also through two opposing points P_1 and P_2 on the ellipse, such that the distance between P_1 and P_2 is maximum. The *minor semi-axis* is half of a line segment passing through the center and also through two opposing points Q_1 and Q_2 on the ellipse, such that the distance between Q_1 and Q_2 is minimum. The full line segments, representing the "long diameter" and the "short diameter," are called the *major axis* and the *minor axis*, respectively.

Let E be an ellipse whose major semi-axis has a length of r_1 and whose minor semi-axis has a length of r_2, as shown in Fig. 8-7. The interior area, A, of this ellipse is given by:

$$A = \pi r_1 r_2$$

PROBLEM 8-6

Find the circumference of a circle that has a diameter of 4.252 m.

SOLUTION 8-6

Use the formula for finding the circumference of a circle in terms of its diameter, and "plug in" the value $d = 4.252$, as follows:

$$B = \pi d$$
$$= 3.14159 \times 4.252$$
$$= 13.36 \text{ m}$$

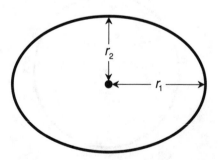

Fig. 8-7. Dimensions of an ellipse.

PROBLEM 8-7
Find the interior area of a circle that has a diameter of 4.2521 m.

SOLUTION 8-7
Use the formula for finding the interior area of a circle in terms of its diameter, and "plug in" the value $d = 4.2521$, as follows:

$$A = \pi d^2/4$$
$$= 3.14159 \times (4.2521)^2/4$$
$$= 14.200 \text{ m}^2$$

PROBLEM 8-8
Find the interior area of an ellipse whose semi-axes measure 2.35 m and 1.1468 m.

SOLUTION 8-8
Use the formula for the interior area of an ellipse in terms of the lengths of its semi-axes, letting $r_1 = 2.35$ and $r_2 = 1.1468$, as follows:

$$A = \pi r_1 r_2$$
$$= 3.14159 \times 2.35 \times 1.1468$$
$$= 8.47 \text{ m}^2$$

Other Formulas

Here are a few more formulas describing some geometric scenarios that you can expect to encounter in the real world.

PERIMETER OF REGULAR POLYGON INSCRIBED IN CIRCLE

Let V be a regular plane polygon having n sides, and whose vertices P_1, P_2, P_3, ..., P_n lie on a circle of radius r (Fig. 8-8). The perimeter, B, of the polygon is given by the following formula:

$$B = 2nr \sin (180/n)^\circ$$

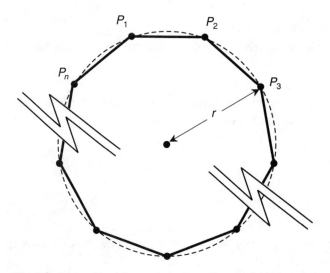

Fig. 8-8. Dimensions of a regular polygon inscribed within a circle.

INTERIOR AREA OF REGULAR POLYGON INSCRIBED IN CIRCLE

Let V be a regular polygon as defined above and in Fig. 8-8. The interior area, A, of the polygon is given by:

$$A = (nr^2/2) \sin (360/n)°$$

PERIMETER OF REGULAR POLYGON CIRCUMSCRIBING CIRCLE

Suppose V is a regular plane polygon having n sides whose center points P_1, P_2, $P_3, \ldots, P_n$ lie on a circle of radius r (Fig. 8-9). The perimeter, B, of the polygon is given by the following formula:

$$B = 2nr \tan (180/n)°$$

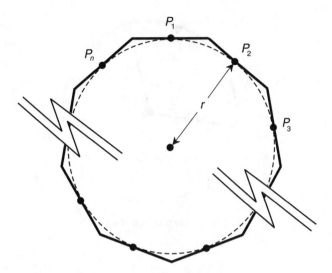

Fig. 8-9. Dimensions of a regular polygon that circumscribes a circle.

INTERIOR AREA OF REGULAR POLYGON CIRCUMSCRIBING CIRCLE

Let V be a regular polygon as defined above and in Fig. 8-9. The interior area, A, of the polygon is given by:

$$A = nr^2 \tan (180/n)°$$

PERIMETER OF CIRCULAR SECTOR

Let S be a sector of a circle whose radius is r (Fig. 8-10). Let θ be the measure of the *apex angle* (the angle between the two radial lines that define the extent of the sector), expressed in radians. The perimeter, B, of the sector is given by the following formula:

$$B = r(2 + \theta)$$

If the apex angle θ is given in degrees rather than in radians, the following formula applies:

$$B = r(2 + \theta \pi / 180)$$

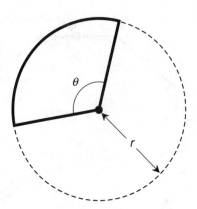

Fig. 8-10. Dimensions of a circular sector.

INTERIOR AREA OF CIRCULAR SECTOR

Let S be a sector of a circle as defined above and in Fig. 8-10. The interior area, A, of the sector is given by:

$$A = r^2\theta/2$$

If the apex angle θ is specified in degrees rather than in radians, use this formula:

$$A = r^2\theta\pi/360$$

A NOTE ABOUT "EXACT VALUES"

Here's a little tidbit that was mentioned in Chapter 3, but bears repeating. In some of the formulas presented in this chapter (and throughout this book), there are defined constants. In the four equations immediately above, you will see constants of 2, 180, π, and 360. When you want to decide on the number of justifiable significant figures in the outcome of a calculation, the values of constants of this sort can be considered "exact." They are accurate to as many significant figures as you need. In the case of the constant 2, for example, you can call it 2.00000..., with any number of zeros you want. In the case of π, you can extend its decimal rendition to as many digits as necessary. The same is true for obvious whole-number quantities, such as the number of sides in a polygon. When any two "exact values" are added to each other, subtracted from each other, multiplied by each other, or divided by each other, or when an "exact value" is taken to a whole-number power, the result is another "exact value" for purposes of calculation.

PROBLEM 8-9
Suppose a *regular decagon* (10-sided polygon) is inscribed within a circle whose radius is 53.25 cm. What is the perimeter of this decagon in meters?

SOLUTION 8-9
Use the formula for the perimeter of a regular polygon inscribed within a circle, setting $n = 10$ (exactly!) and $r = 53.25$. The perimeter in centimeters is found using this formula:

$$B = 2nr \sin (180/n)°$$
$$= 2 \times 10 \times 53.25 \times \sin (180/10)°$$
$$= 1065 \times \sin 18°$$
$$= 1065 \times 0.309017$$
$$= 329.1 \text{ cm}$$

Because 1 m = 100 cm, this is equal to 3.291 m.

PROBLEM 8-10
What is the interior area of the decagon described in Problem 8-9, in square meters?

SOLUTION 8-10
Use the formula for the interior area of a regular polygon inscribed within a circle, setting $n = 10$ and $r = 53.25$. The interior area in square centimeters (cm^2) is found using this formula:

$$A = (nr^2/2) \sin (360/n)°$$
$$= 10 \times (53.25^2/2) \times \sin (360/10)°$$
$$= 14{,}178 \times \sin 36°$$
$$= 14{,}178 \times 0.58779$$
$$= 8333.7 \text{ cm}^2$$
$$= 8.334 \times 10^3 \text{ cm}^2$$

Because 1 m^2 = 10^4 cm^2, this is equal to 0.8334 m^2.

Quick Practice

Here are some practice problems that cover the material presented in this chapter. Solutions follow the problems.

PROBLEMS

1. Find the interior area, in square meters, of an equilateral triangle measuring 2.000 m on each side.

2. Find the interior area, in square centimeters, of a right triangle having sides measuring 300 mm, 400 mm, and 500 mm.

3. Suppose a square has a perimeter in meters that is exactly equal to its interior area in square meters. What is the length of each side in meters? (Do not consider the trivial case, where the length of each side is 0.)

4. If the length of the diagonal of a square is increased by a factor of 5, by what factor does its perimeter increase? By what factor does its interior area increase?

5. Suppose a trapezoid has two parallel sides measuring 4.57 m and 6.03 m. These parallel sides are 1.00 m apart. What is the interior area of the trapezoid in square meters?

SOLUTIONS

1. Use the second formula for the interior area of a triangle given earlier in this chapter. In an equilateral triangle, all three angles measure 60°. Set $s_1 = 2.000$, $s_2 = 2.000$, and $\theta = 60°$ (this angular measure is exact). Then:

$$A = (s_1 s_2 \sin \theta)/2$$
$$= 2.000 \times 2.000 \times (\sin 60°)/2$$
$$= 4.000 \times 0.8660254/2$$
$$= 1.732 \text{ m}^2$$

2. First, convert all lengths to centimeters, because we want to find the area in square centimeters. Consider the base length to be 30.0 cm. Then the height is 40.0 cm, because the angle between the 30.0 cm and 40.0 cm

sides is a right angle. Using the first formula for the interior area of a triangle given earlier in this chapter, set $s_1 = 30.0$ and $h = 40.0$. Then:

$$A = s_1 h / 2$$
$$= 30.0 \times 40.0 / 2$$
$$= 600 \text{ cm}^2$$

3. Let s be the length of each side of this square. Then consider the formulas for perimeter, B, and area, A, of a square, as follows:

$$B = 4s$$
$$A = s^2$$

Because the perimeter and the area are represented by the same number (even though the units differ in terms of dimension), we can set $B = A$, obtaining this single-variable equation:

$$4s = s^2$$

We are told that $s \neq 0$. Therefore, it is all right to divide each side of the above equation by s. This yields the solution directly as $4 = s$, which means that the square we seek has sides that are each exactly 4 m long. The perimeter of such a square is exactly 16 m, and the interior area is exactly 16 m^2.

4. If the length of the diagonal of a square increases by a factor of 5, then the length of each side also increases by a factor of 5. In effect, we magnify the square by a linear factor of 5, because all sides of a square are of equal length. That means the perimeter of the square increases by a factor of 5. However, the interior area of the square increases by a factor of 5^2, or 25. This is true regardless of the initial size of the square.

5. We do not have to know the lengths of the sides connecting the ends of the parallel sides of this trapezoid in order to determine its interior area. Use the formula for the interior area of a trapezoid, letting $s_1 = 4.57$, $s_3 = 6.03$, and $h = 1.00$, as follows:

$$A = (s_1 h + s_3 h)/2$$
$$= (4.57 \times 1.00 + 6.03 \times 1.00)/2$$
$$= (4.57 + 6.03)/2$$
$$= 10.60/2$$
$$= 5.30 \text{ m}^2$$

Quiz

This is an "open book" quiz. You may refer to the text in this chapter. You may draw diagrams if that will help you visualize things. A good score is 8 correct. Answers are in the back of the book.

1. If the radius of a circle is doubled, what happens to the perimeter of a regular decagon (10-sided polygon) inscribed within the circle?

   (a) It increases by a factor of the square root of 2.
   (b) It increases by a factor of 2.
   (c) It increases by a factor of 4.
   (d) We cannot say, unless we know the radius of the circle to begin with.

2. If the radius of a circle is doubled, then what happens to the interior area of a regular decagon (10-sided polygon) inscribed within the circle?

   (a) It increases by a factor of the square root of 2.
   (b) It increases by a factor of 2.
   (c) It increases by a factor of 4.
   (d) We cannot say, unless we know the radius of the circle to begin with.

3. Suppose you have a circle C of radius r. Imagine a sector S of circle C, such that S has an apex angle of 30.00°. Now imagine a second circle D, also of radius r. Imagine a sector T of circle D, that T has an apex angle of 90.00°. How does the perimeter of T compare with the perimeter of S? Express the answer to four significant figures.

   (a) It increases by a factor of 1.415.
   (b) It increases by a factor of 2.830.
   (c) It increases by a factor of 3.000.
   (d) It increases by a factor of 9.000.

4. Suppose you have a circle C of radius r. Imagine a sector S of circle C, such that S has an apex angle of 45.00°. Now imagine a second circle D, of radius $3r$. Imagine a sector T of circle D, that T has an apex angle of 45.00°. How does the interior area of T compare with the interior area of S? Express the answer as a number to four significant figures.

   (a) It increases by a factor of 1.415.
   (b) It increases by a factor of 2.830.
   (c) It increases by a factor of 3.000.
   (d) It increases by a factor of 9.000.

5. Imagine two parallel lines, called line L and line M, such that the lines are separated by a distance s. Let P and Q be two points on line L, such that the points are separated by a distance t. Let R be a point on line M. Consider the triangle PQR with vertices at points P, Q, and R. What is a formula for the interior area A of this triangle?

 (a) There is no way to state this formula without more information.
 (b) $A = st$.
 (c) $A = 2st$.
 (d) $A = st/2$.

6. Imagine two parallel lines, called line L and line M, such that the lines are separated by a distance s. Let P and Q be two points on line L, such that the points are separated by a distance t. Let R be a point on line M. Consider the triangle PQR with vertices at points P, Q, and R. What is a formula for the perimeter B of this triangle?

 (a) There is no way to state this formula without more information.
 (b) $B = st$.
 (c) $B = 2st$.
 (d) $B = st/2$.

7. What is the perimeter of a rhombus if all four sides are 4530 mm long?

 (a) 18.12 m.
 (b) 10.26 m.
 (c) 20.52 m.
 (d) More information is needed in order to calculate this.

8. What is the interior area of a rhombus if all four sides are 4530 mm long?

 (a) 18.12 m^2.
 (b) 10.26 m^2.
 (c) 20.52 m^2.
 (d) More information is needed in order to calculate this.

9. Suppose you are 1533 mm tall. If you stand in the sunshine on a flat, level surface and your shadow is 2044 mm long, what is the distance from the top of your head to the end of your shadow?

 (a) 1.770 m.
 (b) 3.789 m.
 (c) 2.555 m.
 (d) More information is needed in order to calculate this.

10. Suppose a triangle has two sides with lengths of 24.00 m and 10.00 m, and the interior area of the triangle is 120.0 m^2. What is the length of the third side?

    (a)　26.00 m.

    (b)　28.00 m.

    (c)　30.00 m.

    (d)　More information is needed in order to calculate this.

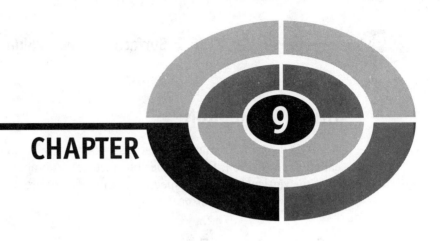

Surface Area and Volume in Three Dimensions

In this chapter, you'll learn how to find the surface areas and volumes of various simple geometric solids in three dimensions (3D).

Straight-Edged Objects

In Euclidean (that is, "non-warped") three-space, geometric solids with straight edges have flat faces, also called *facets*, each of which is a plane polygon. An object of this sort is known as a *polyhedron*.

THE TETRAHEDRON

A polyhedron in 3D must have at least four faces. A four-faced polyhedron is called a *tetrahedron*. Each of the four faces of a tetrahedron is a triangle. There are four vertices. Any four specific points, if they are not all in a single plane, define a unique tetrahedron.

Surface Area of Tetrahedron

Figure 9-1 shows a tetrahedron. The surface area is found by adding up the interior areas of all four triangular faces. In the case of a *regular tetrahedron*, all six edges have the same length, and therefore all four faces are equilateral triangles. If the length of each edge of a regular tetrahedron is equal to s units, then the surface area, B, of the whole four-faced regular tetrahedron, is given by:

$$B = 3^{1/2} s^2$$

where $3^{1/2}$ represents the square root of 3, or approximately 1.732. This also happens to be twice the sine of 60°, which is the angle between any two edges of the figure.

Volume of Tetrahedron

Imagine a tetrahedron whose base is a triangle with area A, and whose height is h as shown in Fig. 9-1. The volume, V, of the tetrahedron is given by this formula:

$$V = Ah/3$$

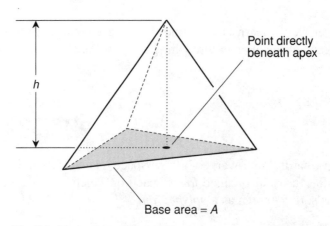

Point directly beneath apex

h

Base area $= A$

Fig. 9-1. Dimensions of a tetrahedron.

RECTANGULAR PYRAMID

Figure 9-2 illustrates a *rectangular pyramid*. This figure has a rectangular base and four slanted faces. If the base is a square, and if the apex (the top of the pyramid) lies directly above a point at the center of the base, then the figure is a *symmetrical square pyramid*, and all of the slanted faces are isosceles triangles.

Surface Area of Symmetrical Square Pyramid

In the case of a symmetrical square pyramid where the length of each slanted edge, called the *slant height*, is equal to s units and the length of each edge of the base is equal to t units, the surface area, B, is given by:

$$B = t^2 + 2t \, (s^2 - t^2/4)^{1/2}$$

Volume of Rectangular Pyramid

Imagine a rectangular pyramid whose base has area A, and whose height is h, as shown in Fig. 9-2. The volume, V, of the pyramid is given by:

$$V = Ah/3$$

This formula works even if the apex is not directly above the center of the base. In fact, it works even if the apex is above a point outside the base. The height, h, is always defined as the distance between the apex and the plane containing the base, as measured along a line normal (perpendicular) to the plane containing the base.

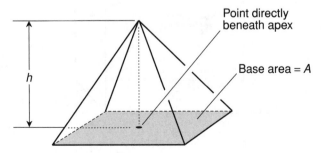

Fig. 9-2. Dimensions of a rectangular pyramid.

THE CUBE

Figure 9-3 illustrates a *cube*. This is a *regular hexahedron* (six-sided polyhedron). It has 12 edges, each of which is of the same length. Each of the six faces is a square.

Surface Area of Cube

Imagine a cube whose edges each have length *s*, as shown in Fig. 9-3. The surface area, *A*, of the cube is given by:

$$A = 6s^2$$

Volume of Cube

Imagine a cube as defined above and in Fig. 9-3. The volume, *V*, is given by:

$$V = s^3$$

THE RECTANGULAR PRISM

Figure 9-4 illustrates a *rectangular prism*. This is a hexahedron, such that all six faces are rectangles. At each vertex, the edges converge at mutual right angles. The figure has 12 edges, but they are not necessarily all the same length.

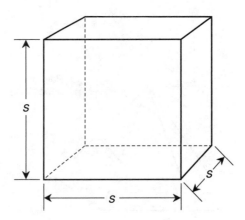

Fig. 9-3. Dimensions of a cube.

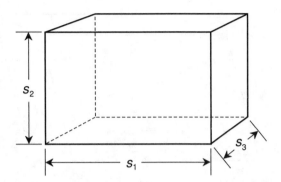

Fig. 9-4. Dimensions of a rectangular prism.

Surface Area of Rectangular Prism

Imagine a rectangular prism whose edges have lengths s_1, s_2, and s_3, as shown in Fig. 9-4. The surface area, A, of the prism is given by:

$$A = 2s_1s_2 + 2s_1s_3 + 2s_2s_3$$

Volume of Rectangular Prism

Imagine a rectangular prism as defined above and in Fig. 9-4. The volume, V, is given by:

$$V = s_1s_2s_3$$

THE PARALLELEPIPED

A *parallelepiped* is a six-faced polyhedron in which each face is a parallelogram, and opposite pairs of faces are *congruent* (meaning that they have identical size and shape). The figure has 12 edges. An example is shown in Fig. 9-5.

Surface Area of Parallelepiped

Consider a parallelepiped with faces of lengths s_1, s_2, and s_3. Suppose the acute angles between pairs of edges are x, y, and z, as shown in Fig. 9-5. The surface area, A, of the parallelepiped is given by the following formula:

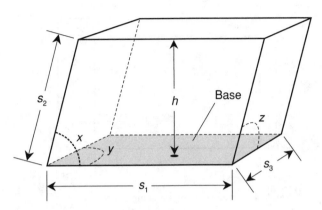

Fig. 9-5. Dimensions of a parallelepiped.

$$A = 2s_1s_2 \sin x + 2s_1s_3 \sin y + 2s_2s_3 \sin z$$

where $\sin x$ represents the sine of angle x, $\sin y$ represents the sine of angle y, and $\sin z$ represents the sine of angle z.

Volume of Parallelepiped

Imagine a parallelepiped whose faces have lengths s_1, s_2, and s_3, and that has acute angles between edges of x, y, and z, as shown in Fig. 9-5. Suppose further that the height of the parallelepiped, as measured along a line normal to the plane containing the base, is equal to h. The volume, V, of the enclosed solid is equal to the product of the base area and the height:

$$V = hs_1s_3 \sin y$$

PROBLEM 9-1

Suppose you want to paint the interior walls of a room in a house. The room is shaped like a rectangular prism. The ceiling is 3.000 m above the floor. The floor and the ceiling both measure 4.200 m by 5.500 m. There are two windows, the outer frames of which both measure 1.500 m high by 1.000 m wide. There is one doorway, the outer frame of which measures 2.500 m high by 1.000 m wide. One liter of paint can be expected to cover exactly 20.00 m² of wall area. How much paint, in liters, will you need to completely do the job?

SOLUTION 9-1

It is necessary to find the amount of wall area that this room has. Based on the information given, we can conclude that the rectangular prism formed by the edges between walls, floor, and ceiling measures 3.000 m high by 4.200 m wide by 5.500 m deep. Let $s_1 = 3.000$, $s_2 = 4.200$, and $s_3 = 5.500$ (with all units assumed to be in meters) to find the surface area A of the rectangular prism, in square meters, neglecting the area subtracted by the windows and doorway. Using the formula:

$$A = 2s_1s_2 + 2s_1s_3 + 2s_2s_3$$
$$= (2 \times 3.000 \times 4.200) + (2 \times 3.000 \times 5.500) + (2 \times 4.200 \times 5.500)$$
$$= 25.20 + 33.00 + 46.20$$
$$= 104.40 \text{ m}^2$$

There are two windows measuring 1.500 m by 1.000 m; each of these therefore takes away $1.500 \times 1.000 = 1.500 \text{ m}^2$ of area. The doorway measures 2.500 m by 1.000 m, so it takes away $2.500 \times 1.000 = 2.500 \text{ m}^2$. Thus the windows and doorway combined take away $1.500 + 1.500 + 2.500 = 5.500 \text{ m}^2$ of wall space. We must also take away the combined areas of the floor and ceiling, represented by $2s_2s_3 = 46.20$. The wall area to be painted, call it A_w, is therefore:

$$A_w = (104.40 - 5.500) - 46.20$$
$$= 52.70 \text{ m}^2$$

We are told that a liter of paint can cover 20.00 m^2. Therefore, we will need 52.70/20.00, or 2.635, liters of paint to do this job.

Cones and Cylinders

A *cone* has a circular or elliptical base and an apex point. The cone itself consists of the union of the following sets of points:

- The circle or ellipse itself.
- All points inside, and in the plane determined by, the circle or ellipse.
- All line segments connecting the circle or ellipse (not including its interior) and the apex point.

The interior of the cone consists of the set of all points within the cone.

A *cylinder* has a circular or elliptical base, and a circular or elliptical top that is congruent to the base and that lies in a plane parallel to the base. The cylinder itself consists of the union of the following sets of points:

- The base.
- All points inside, and in the plane determined by, the base.
- The top.
- All points inside, and in the plane determined by, the top.
- All line segments connecting corresponding points on the base and the top (not including their interiors).

The interior of the cylinder consists of the set of all points within the cylinder. These are general definitions, and they encompass a great variety of objects! In this chapter, we'll look only at cones and cylinders whose bases are circles.

THE RIGHT CIRCULAR CONE

A *right circular cone* has a base that is a circle, and an apex point that lies on a line normal (perpendicular) to the plane of the base, and that passes through the center of the base. An example is shown in Fig. 9-6.

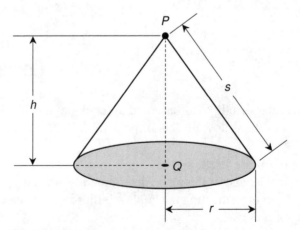

Fig. 9-6. Dimensions of a right circular cone.

Surface Area of Right Circular Cone

Imagine a right circular cone as shown in Fig. 9-6. Let P be the apex of the cone, and let Q be the center of the base. Let r be the radius of the base, let h be the height of the cone (the length of line segment PQ), and let s be the slant height of the cone as measured from any point on the edge of the base to the apex P. The surface area S_1 of the cone, including the base, is given by either of the following formulas:

$$S_1 = \pi r^2 + \pi rs$$
$$S_1 = \pi r^2 + \pi r(r^2 + h^2)^{1/2}$$

The surface area S_2 of the cone, not including the base, is called the *lateral surface area* and is given by either of the following:

$$S_2 = \pi rs$$
$$S_2 = \pi r(r^2 + h^2)^{1/2}$$

Volume of Right Circular Cone

Imagine a right circular cone as defined above and in Fig. 9-6. The volume, V, of the interior of the cone is given by this formula:

$$V = \pi r^2 h/3$$

Surface Area of Frustum of Right Circular Cone

Imagine a right circular cone that is *truncated* (cut off) by a plane parallel to the base. This is called a *frustum* of the right circular cone. Let P be the center of the circle defined by the truncation, and let Q be the center of the base, as shown in Fig. 9-7. Suppose line segment PQ is perpendicular to the base. Let r_1 be the radius of the top, let r_2 be the radius of the base, let h be the height of the object (the length of line segment PQ), and let s be the slant height. Then the surface area S_1 of the object (including the base and the top) is given by either of the following formulas:

$$S_1 = \pi(r_1 + r_2)[h^2 + (r_2 - r_1)^2]^{1/2} + \pi(r_1^2 + r_2^2)$$
$$S_1 = \pi s(r_1 + r_2) + \pi(r_1^2 + r_2^2)$$

The surface area S_2 of the object (not including the base or the top) is given by either of the following:

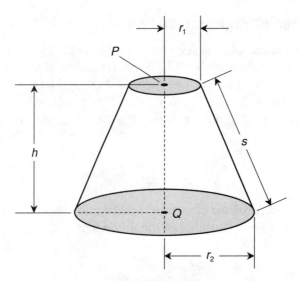

Fig. 9-7. Dimensions of a frustum of a right circular cone.

$$S_2 = \pi(r_1 + r_2)[h^2 + (r_2 - r_1)^2]^{1/2}$$
$$S_2 = \pi s(r_1 + r_2)$$

Volume of Frustum of Right Circular Cone

Imagine a frustum of a right circular cone as defined above and in Fig. 9-7. The volume, V, of the interior of the object is given by this formula:

$$V = \pi h(r_1^2 + r_1 r_2 + r_2^2)/3$$

THE SLANT CIRCULAR CONE

A *slant circular cone* has a base that is a circle, and an apex point such that a normal line from the apex point to the plane of the base does not pass through the center of the base. In some cases (such as the example shown in Fig. 9-8), this normal line does not pass through the base.

Volume of Slant Circular Cone

Imagine a cone whose base is a circle. Let P be the apex of the cone, and let Q be a point in the plane X containing the base such that line segment PQ is

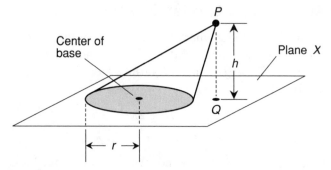

Fig. 9-8. Dimensions of a slant circular cone.

perpendicular to X, as shown in Fig. 9-8. Let h be the height of the cone (the length of line segment PQ). Let r be the radius of the base. Then the volume, V, of the corresponding cone is given by:

$$V = \pi r^2 h / 3$$

THE RIGHT CIRCULAR CYLINDER

A *right circular cylinder* has a circular base and a circular top. The base and the top lie in parallel planes. The center of the base and the center of the top lie along a line that is normal to both the plane containing the base and the plane containing the top (Fig. 9-9).

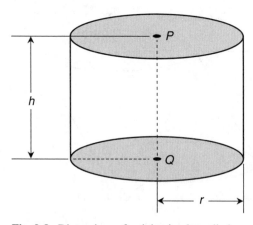

Fig. 9-9. Dimensions of a right circular cylinder.

Surface Area of Right Circular Cylinder

Imagine a right circular cylinder where P is the center of the top and Q is the center of the base (Fig. 9-9). Let r be the radius of the cylinder, and let h be the height (the length of line segment PQ). Then the surface area S_1 of the cylinder, including the base, is given by:

$$S_1 = 2\pi rh + 2\pi r^2$$
$$= 2\pi r(h + r)$$

The lateral surface area S_2 of the cylinder (not including the base) is given by:

$$S_2 = 2\pi rh$$

Volume of Right Circular Cylinder

Imagine a right circular cylinder as defined above and shown in Fig. 9-9. The volume, V, of the cylinder is given by:

$$V = \pi r^2 h$$

THE SLANT CIRCULAR CYLINDER

A *slant circular cylinder* has a circular base and a circular top. The base and the top lie in parallel planes. The center of the base and the center of the top lie along a line that is not perpendicular to the planes that contain them (Fig. 9-10).

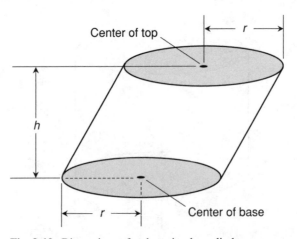

Fig. 9-10. Dimensions of a slant circular cylinder.

Volume of Slant Circular Cylinder

Imagine a slant circular cylinder as defined above and in Fig. 9-10. The volume, V, of the cylinder is given by the same formula as that for the right circular cylinder:

$$V = \pi r^2 h$$

PROBLEM 9-2

Suppose a cylindrical water tower is exactly 30 m high and exactly 10 m in radius. How many liters of water can it hold, assuming the entire interior can be filled with water? (One liter is equal to a cubic decimeter, or the volume of a cube measuring 0.1 m on an edge.) Round the answer off to the nearest 10 liters.

SOLUTION 9-2

Use the formula for the volume of a right circular cylinder to find the volume in cubic meters:

$$V = \pi r^2 h$$

Plugging in the numbers, let $r = 10$, $h = 30$, and $\pi = 3.14159$:

$$V = 3.14159 \times 10^2 \times 30$$
$$= 3.14159 \times 100 \times 30$$
$$= 9424.77$$

One liter is the volume of a cube measuring 10 cm, or 0.1 m, on an edge. Thus, there are 1000 liters in a cubic meter. This means that the amount of water the tower can hold, in liters, is equal to 9424.77 × 1000, or 9,424,770.

PROBLEM 9-3

Imagine a circus tent shaped like a right circular cone. Suppose its diameter is exactly 50 m, and the height at the center is exactly 20 m. How much canvas has been used to make the tent? Express the answer rounded up to the next higher square meter. (Assume the floor of the tent is bare ground.)

SOLUTION 9-3

Use the formula for the lateral surface area, S, of the right circular cone.

$$S = \pi r (r^2 + h^2)^{1/2}$$

We know that the diameter is 50 m, so the radius is 25 m. Therefore, $r = 25$. We also know that $h = 20$. Let $\pi = 3.14159$. Then:

$$S = 3.14159 \times 25 \times (25^2 + 20^2)^{1/2}$$
$$= 3.14159 \times 25 \times (625 + 400)^{1/2}$$
$$= 3.14159 \times 25 \times 1025^{1/2}$$
$$= 3.14159 \times 25 \times 32.0156$$
$$= 2514.49$$

There are 2515 m^2 of canvas, rounded up to the next higher square meter.

Other Solids

There are many types of geometric solids with curved surfaces throughout. Here, we examine three of the most common: the *sphere*, the *ellipsoid*, and the *torus*.

THE SPHERE

Consider a specific point P in 3D space. The surface of a sphere S consists of the set of all points at a specific distance or radius r from point P. The interior of sphere S, including the surface, consists of the set of all points whose distance from point P is less than or equal to r. The interior of sphere S, not including the surface, consists of the set of all points whose distance from P is less than r.

Surface Area of Sphere

Imagine a sphere S having radius r as shown in Fig. 9-11. The surface area, A, of the sphere is given by:

$$A = 4\pi r^2$$

Volume of Sphere

Imagine a sphere S as defined above and in Fig. 9-11. The volume, V, of the solid enclosed by the sphere is given by:

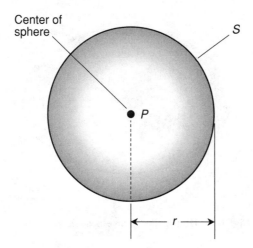

Fig. 9-11. Dimensions of a sphere.

$$V = 4\pi r^3/3$$

This volume applies to the interior of sphere S, either including the surface or not including it, because the surface has zero volume.

THE ELLIPSOID

Let E be a set of points that forms a closed surface. Then E is an ellipsoid if and only if, for any plane X that intersects E, the intersection between E and X is either a single point, a circle, or an ellipse. Figure 9-12 shows an ellipsoid E with center point P and radii (also called semi-axes) measuring r_1, r_2, and r_3, as specified in a 3D rectangular coordinate system with P at the origin. If r_1, r_2, and r_3 are all equal, then E is a sphere, which is a special case of the ellipsoid.

Volume of Ellipsoid

Imagine an ellipsoid whose radii are r_1, r_2, and r_3 (Fig. 9-12). The volume, V, of the enclosed solid is given by:

$$V = 4\pi r_1 r_2 r_3/3$$

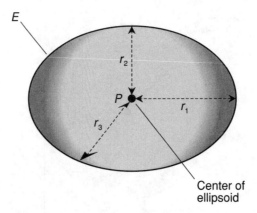

Fig. 9-12. Dimensions of an ellipsoid.

THE TORUS

Imagine a ray PQ, and a small circle C centered on point Q whose radius is less than half of the distance between points P and Q. Suppose ray PQ, along with the small circle C centered at point Q, is rotated around its end point, P, so that point Q describes a large circle that lies in a plane perpendicular to the plane containing the small circle C. The resulting set of points in 3D space, "traced out" by circle C, is a torus. Figure 9-13 shows a torus T thus constructed, with center point P. The *inner radius* is r_1 and the *outer radius* is r_2.

Surface Area of Torus

Imagine a torus with an inner radius of r_1 and an outer radius of r_2 as shown in Fig. 9-13. The surface area, B, of the torus is given by:

$$B = \pi^2(r_2 + r_1)(r_2 - r_1)$$

Volume of Torus

Let T be a torus as defined above and in Fig. 9-13. The volume, V, of the enclosed solid is given by:

$$V = \pi^2(r_2 + r_1)(r_2 - r_1)^2/4$$

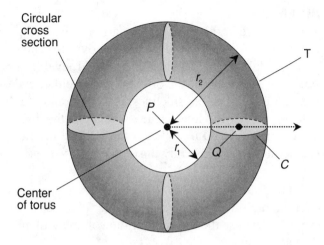

Fig. 9-13. Dimensions of a torus.

PROBLEM 9-4
Suppose a football field is to be covered by an inflatable dome that takes the shape of a half-sphere. If the radius of the dome is 100 m, what is the volume of air enclosed by the dome in cubic meters? Find the result to the nearest 1000 cubic meters.

SOLUTION 9-4
First, find the volume V of a sphere whose radius is 100 m, and then divide the result by 2. Let $\pi = 3.14159$. Using the formula with $r = 100$ gives this result:

$$V = 4\pi r^3/3$$
$$= (4 \times 3.14159 \times 100^3)/3$$
$$= (4 \times 3.14159 \times 1,000,000)/3$$
$$= 4,188,787$$

Therefore, $V/2 = 4,188,787/2 = 2,094,393.5$. Rounding off to the nearest 1000 cubic meters, we get 2,094,000 m^3 as the volume of air enclosed by the dome.

PROBLEM 9-5

Suppose the dome in the previous scenario is a half-ellipsoid. Imagine that the height of the ellipsoid is 70 m above its center point, which lies in the middle of the 50-yard line at field level. Suppose that the distance from the center of the 50-yard line to either end of the dome, as measured parallel to the sidelines, is 120 m, and the distance from the center of the 50-yard line, as measured along the line containing the 50-yard line itself, is 90 m. What is the volume of air, to the nearest 1000 cubic meters, enclosed by this dome?

SOLUTION 9-5

Consider the radii with respect to the center point, as $r_1 = 120$, $r_2 = 90$, and $r_3 = 70$. Then use the formula for the volume V of an ellipsoid, and calculate as follows:

$$\begin{aligned} V &= 4\pi r_1 r_2 r_3 / 3 \\ &= (4 \times 3.14159 \times 120 \times 90 \times 70)/3 \\ &= (4 \times 3.14159 \times 756{,}000)/3 \\ &= 3{,}166{,}723 \end{aligned}$$

Therefore, $V/2 = 3{,}166{,}723/2 = 1{,}583{,}361.5$. Rounding off to the nearest 1000 cubic meters, we get 1,583,000 m^3 as the volume of air enclosed by the half-ellipsoidal dome.

Quick Practice

Here are some practice problems that cover the material presented in this chapter. Solutions follow the problems.

PROBLEMS

1. What is the surface area, in square meters, of a regular tetrahedron that measures 1.000 m on each edge?

2. What is the volume, in cubic meters, of a regular tetrahedron that measures exactly 1.635 m on each edge around the base, and whose height is 2.761 m?

3. What is the surface area, in square meters, of a symmetrical square pyramid that measures exactly 2.000 m on each edge around the base, and whose slant height is also 2.000 m?

4. What is the volume, in cubic meters, of a symmetrical square pyramid that measures exactly 2.000 m on each edge around the base, and whose height is 3.000 m?

5. Imagine a room shaped like a rectangular prism that measures 10.00 ft wide by 15.00 ft long, and whose ceiling is 8.000 ft above the floor. How many cubic feet of air are inside this room if it is totally empty (except for air)?

SOLUTIONS

1. Let the length of each edge be $s = 1.000$. Using the formula for the surface area B of a regular tetrahedron, calculate as follows:

$$B = 3^{1/2} s^2$$
$$= 3^{1/2} \times 1.000^2$$
$$= 3^{1/2}$$
$$= 1.732 \text{ m}^2$$

2. First, find the area of the base. Note that the base is an equilateral triangle whose sides all have length $s = 1.635$ m, and whose apex angles all measure $\theta = 60°$. Using the formula for the interior area, A, of an equilateral triangle from the last chapter, and letting $s_1 = s_2 = s$, proceed as follows:

$$A = (s_1 s_2 \sin \theta)/2$$
$$= (s^2 \sin 60°)/2$$
$$= 1.635^2 \times 0.8660254/2$$
$$= 1.157540 \text{ m}^2$$

Use this value for A in the formula for the volume of the tetrahedron, where $h = 2.761$, as follows:

$$V = Ah/3$$
$$= 1.157540 \times 2.761/3$$
$$= 1.065 \text{ m}^3$$

3. In the formula for the surface area, B, of a symmetrical square pyramid, set $s = t = 2.000$. Then calculate B as follows:

$$B = t^2 + 2t\,(s^2 - t^2/4)^{1/2}$$
$$= 2.000^2 + 2 \times 2.000\,(2.000^2 - 2.000^2/4)^{1/2}$$
$$= 4.000 + 4.000\,(4.000 - 1.000)^{1/2}$$
$$= 4.000 + 4.000 \times 3^{1/2}$$
$$= 10.93 \text{ m}^2$$

4. First, determine the area of the base. The base is a square with sides of length $s = 2.000$ m. Using the formula for the interior area of a square:

$$A = s^2$$
$$= 2.000^2$$
$$= 4.000 \text{ m}^2$$

The volume, V, of this regular square pyramid with height $h = 3.000$ m is found as follows:

$$V = Ah/3$$
$$= 4.000 \times 3.000/3$$
$$= 4.000 \text{ m}^3$$

5. Use the formula for the volume, V, of a rectangular prism with edges measuring $s_1 = 10.00$ ft, $s_2 = 15.00$ ft, and $s_3 = 8.000$ ft, as follows:

$$V = s_1 s_2 s_3$$
$$= 10.00 \times 15.00 \times 8.000$$
$$= 1200 \text{ ft}^3$$

Quiz

This is an "open book" quiz. You may refer to the text in this chapter. You may draw diagrams if that will help you visualize things. A good score is 8 correct. Answers are in the back of the book.

1. What is the approximate volume of a circular cone whose base has an area of 30 square units and a height of 10 units?

(a) 100 cubic units.

(b) 150 cubic units.

(c) 300 cubic units.

(d) There is not enough information given here to calculate it.

2. Consider the earth to be a perfect sphere that measures 12,800 km in diameter. What is the approximate surface area of the earth based on this figure?

(a) 5.15×10^8 km^2.

(b) 2.06×10^9 km^2.

(c) 1.10×10^{12} km^2.

(d) 8.80×10^{12} km^2.

3. What is the set of points representing the intersection of a torus with a plane containing its center?

(a) A pair of nonconcentric circles.

(b) A pair of concentric circles.

(c) A pair of nonconcentric ellipses.

(d) To answer this, we must know the orientation of the plane.

4. A rectangular prism has

(a) six edges, all of which are the same length.

(b) eight edges of various lengths.

(c) six faces, all of which are the same shape.

(d) None of the above

5. If all other factors are held constant, the surface area of a torus depends on all of the following except

(a) its inner radius.

(b) its outer radius.

(c) the difference between the squares of its inner and outer radii.

(d) its orientation in space.

6. If all other factors are held constant, the volume of a parallelepiped depends on

(a) the height.

(b) the width.

(c) the ratio of the height to the width.

(d) More than one of the above

7. In a slant circular cylinder, the height is equal to

    (a) the radius of either the base or the top multiplied by 2π.
    (b) the distance between the planes containing the base and the top.
    (c) the distance between the center of the base and the center of the top.
    (d) the distance along any straight line in the periphery.

8. The volume of a rectangular prism is equal to

    (a) the sum of the lengths of its edges.
    (b) the product of the lengths of its edges.
    (c) the sum of the surface areas of its faces.
    (d) the product of the surface areas of its faces.

9. The faces of a tetrahedron

    (a) are all triangles.
    (b) are all quadrilaterals.
    (c) are all congruent to each other.
    (d) all lie in the same plane.

10. Imagine a cube with edges measuring 10.00 m each. Suppose a pyramid is carved from this cube, such that the base of the pyramid corresponds to one of the faces of the cube, and the apex of the pyramid is at the center of the face of the cube opposite the base. What is the approximate volume of the pyramid?

    (a) 1000 m^3.
    (b) 7071 m^3.
    (c) 333.3 m^3.
    (d) There is not enough information given here to calculate it.

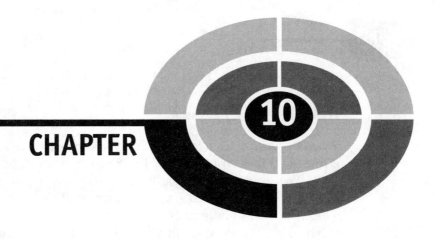

Boolean Algebra

Boolean algebra is a system of *propositional logic* using the binary numbers 1 and 0. These logic states are also called *high* and *low*, respectively. Sometimes they are considered as the equivalent of *true* and *false*. Boolean operations and relations behave differently than their counterparts in real-number and complex-number algebra.

Operations, Relations, and Symbols

In Boolean algebra, statements (or sentences) are represented by uppercase letters of the alphabet. For example, you say, "It is raining outside," and represent this by the letter R. Someone else adds, "It's cold outside," and represents this by the letter C. A third person says, "The weather forecast calls for snow tomorrow," and represents this by the letter S. Still another person claims, "Tomorrow's forecast calls for bright sunshine," and represents this by the letter B.

NEGATION (−)

When we write down a letter to stand for a sentence, we assert that the sentence is true. So if John writes down C in the above situation, he means "It is cold outside." You would not necessarily agree with this if you grew up in Fairbanks, Alaska. You could say, "It's not cold outside." This would be denoted as C with a *negation* symbol in front of it.

There are several ways in which negation, also called the *NOT operation*, can be symbolized. In Boolean algebra, the minus sign (−) is used. Thus, the sentence "It's not cold outside" is denoted −C. In propositional logic, a common symbol for negation is a drooping minus sign (¬). Some texts use a tilde (~) to represent negation. Some put a line over the letter representing the sentence.

Suppose someone comes along and says, "You are correct to say −C. In fact, I'd say it's hot outside!" Suppose this is symbolized H. Does H mean the same thing as −C? Not necessarily. You've seen days that were neither cold nor hot. There can be in-between states such as "cool" (K), "mild" (M), and "warm" (W). But there is no in-between condition when it comes to C and −C. In propositional logic, either it is cold, or else it is not cold. Either it's hot, or else it is not hot. A proposition is either true, or else it is false (not true).

There are logical systems in which in-between states exist. These go by nicknames such as *fuzzy logic*. We won't get involved with them.

LOGICAL CONJUNCTION (×)

Propositional logic doesn't get involved with how the phrases inside a sentence affect each other, but it is concerned with the ways in which distinct, complete sentences interact. Logical statements or *Boolean variables* can be combined to make larger logical structures, called *compound statements* or *Boolean expressions*. The truth or falsity of a Boolean expression depends on the truth or falsity of its components, and on the ways those components are connected.

Suppose someone says, "It's cold outside, and it's raining outside." Using the symbols above, we can write this as:

C AND R

In Boolean algebra, a multiplication symbol (×) is used in place of the word AND. Thus, the expression above is written like this in Boolean algebra:

C × R

Various other symbols are used by mathematicians and logicians for the AND operation, including the ampersand (&), the inverted wedge (∧), the asterisk (*), the period or radix point (.), and the raised dot (·).

The formal term for the *AND operation* is *logical conjunction*. A Boolean expression containing one or more conjunctions has value 1 when, but only when, both or all of its components have value 1. If any of the components have value 0, then the whole expression has value 0.

LOGICAL DISJUNCTION (+)

Suppose that the above conversation about the weather continues. One of your friends says, "It's cold and raining; there is no doubt about that. I have been listening to the radio, and I heard the weather forecast for tomorrow. It's supposed to be colder tomorrow than it is today. But it's going to stay wet. So it might snow tomorrow."

You say, "It will rain or it will snow tomorrow, depending on the temperature."

Your friend says, "It might be a mix of rain and snow together, if the temperature is near freezing."

"So we might get rain, we might get snow, and we might get both," you say.

"Correct. But the weather experts say we are certain to get precipitation of some sort," your friend says. "Water is going to fall from the sky tomorrow. Maybe it will be liquid, maybe it will be solid, and maybe it will be both."

In this case, suppose we let R represent the sentence "It will rain tomorrow," and we let S represent the sentence "It will snow tomorrow." Then we can say:

S OR R

This is an example of *logical disjunction*. In Boolean algebra, the addition symbol is used to represent this *OR operation*. We can now write:

S + R

In some texts, you'll see a wedge (∨) used to represent logical disjunction. That symbol is used occasionally by mathematicians and logicians.

A Boolean expression in which both, or all, of the components are joined by disjunctions has value 1 when, but only when, at least one of the components has value 1. A Boolean expression made up of disjunctions has value 0 when, but only when, all the components have value 0.

Logical disjunction, as we define it here, is the *inclusive OR* operation. There's another logic operation called *exclusive OR*, in which the Boolean expression has value 1 if and only if one, but only one, of the variables has value 1.

(Usually there are only two variables in this type of scenario, so it is sometimes called "either/or.") The exclusive OR operation, abbreviated XOR, is important when logic is applied in engineering, especially in digital electronic circuit design.

LOGICAL IMPLICATION (⇒)

Imagine that the conversation about the weather continues. You and your friends are trying to decide if you should get ready for a snowy day tomorrow, or whether rain and gloom is all you'll have to contend with.

"Does the weather forecast say anything about snow?" you ask.

"Not exactly," one your friends says. "The radio announcer made this statement: 'There is going to be precipitation through tomorrow night, and it's going to get colder tomorrow.' I looked at my car thermometer as she said that, and it said the outdoor temperature was just a little bit above freezing."

"If there is precipitation, and if it gets colder, then it will snow," you say.

"Of course."

"Unless we get an ice storm."

"That won't happen."

"Okay," you say. "Let's discount the ice storm scenario. That means that if there is precipitation tomorrow, and if it is colder tomorrow than it is today, then it will snow tomorrow." (This is a clumsy way to talk, but we're learning about Boolean algebra, not the art of elegant conversation!)

Suppose you use P to represent the sentence, "There will be precipitation tomorrow." In addition, let S represent the sentence "It will snow tomorrow," and let C represent the sentence "It will be colder tomorrow." Then in the above conversation, you have made a compound statement consisting of three sentences, like this:

$$\text{IF (P AND C), THEN S}$$

Another way to write this is:

$$\text{(P AND C) IMPLIES S}$$

In logic or Boolean algebra, the statement "X implies Y" is the equivalent of saying "If X, then Y." There is no question of likelihood or doubt. In Boolean algebra, "X implies Y" means that if event X occurs, then event Y is inevitable. Symbolically, the above proposition is written this way:

$$(P \times C) \Rightarrow S$$

The double-shafted arrow pointing to the right represents *logical implication*, also known as the *IF/THEN operation*. In a logical implication, the "implying" sentence (to the left of the double-shafted arrow) is called the *antecedent*. In the above example, the antecedent is (P × C). The "implied" sentence (to the right of the double-shafted arrow) is called the *consequent*. In the above example, the consequent is S.

Some texts make use of other symbols for logical implication, including the "hook" or "lazy U opening to the left" (⊃), three dots (∴), and a single-shafted arrow pointing to the right (→).

LOGICAL EQUIVALENCE (=)

Suppose one of your friends keeps on going in the above conversation and says, "If it snows tomorrow, then there will be precipitation and it will be colder."

For a moment you hesitate, because this isn't the way you'd usually think about this kind of situation. But you have to agree, "That is true. It sounds strange, but it's true." Your friend has just made this implication:

$$S \Rightarrow (P \times C)$$

Implication holds in both directions here. You and your friend agree that both of the following implications are valid:

$$(P \times C) \Rightarrow S$$
$$S \Rightarrow (P \times C)$$

These two implications can be combined into a conjunction, because we are asserting them both together:

$$[(P \times C) \Rightarrow S] \times [S \Rightarrow (P \times C)]$$

When an implication is valid in both directions, the situation is defined as a case of *logical equivalence*. The above statement can be rewritten as follows:

$$(P \times C) \text{ IF AND ONLY IF } S$$

Mathematicians sometimes shorten the phrase "if and only if" to the single word "iff." So we can also write:

$$(P \times C) \text{ IFF } S$$

The Boolean symbol for logical equivalence is the equals sign (=). Other symbols can be used. Sometimes you'll see a three-barred equals sign (≡), a

single-shafted, double-headed arrow ($\leftrightarrow$), or a double-shafted, double-headed arrow ($\Leftrightarrow$). Symbolically, for our purposes, the above logical equation becomes:

$$(P \times C) = S$$

PROBLEM 10-1

Give an example of a situation in which logical implication holds in one direction but not in the other.

SOLUTION 10-1

Consider this statement: "If it is overcast, then there are clouds in the sky." This statement is true. Suppose we let O represent "It is overcast" and K represent "There are clouds in the sky." Then we have this, symbolically:

$$O \Rightarrow K$$

If we reverse this, we get a statement that isn't necessarily true. Consider:

$$K \Rightarrow O$$

This translates to, "If there are clouds in the sky, then it is overcast." We have all seen days or nights in which there were clouds in the sky, but it was not overcast.

Truth Tables

A *truth table* is a method of denoting all possible combinations of truth values for the variables in a proposition. The values for the individual variables, with all possible permutations, are shown in vertical columns at the left. The truth values for Boolean expressions, as they are built up from *atomic propositions*, are shown in horizontal rows. In Boolean algebra, an atomic proposition always consists of either logical constant (that is, 0 or 1), or a single variable. Thus, an atomic proposition can't be broken down into smaller logical components; it is like a "fundamental particle of logic."

TRUTH TABLE FOR NEGATION

The simplest truth table is the one for negation, which operates on a single variable. Table 10-1 shows how this works for a variable X.

Table 10-1.   Truth table for negation.

X	–X
0	1
1	0

TABLE FOR CONJUNCTION

Let X and Y be two logical variables. Conjunction (X × Y) produces results as shown in Table 10-2. The resultant of the AND operation has value 1 when, but only when, both variables have value 1. Otherwise, the operation has value 0.

TABLE FOR DISJUNCTION

Logical disjunction for two variables (X + Y) has a truth table that looks like Table 10-3. The resultant of the inclusive OR operation has value 1 when either or both of the variables equal 1. If both of the variables are 0, then the operation has value 0.

TABLE FOR IMPLICATION

A logical implication is valid (that is, it has truth value 1) except when the antecedent has value 1 and the consequent has value 0. Table 10-4 shows the truth values for logical implication.

Table 10-2.   Truth table for conjunction (Boolean multiplication).

X	Y	X × Y
0	0	0
0	1	0
1	0	0
1	1	1

Table 10-3.   Truth table for disjunction (Boolean addition).

X	Y	X + Y
0	0	0
0	1	1
1	0	1
1	1	1

PROBLEM 10-2

Give an example of a logical implication that is obviously invalid.

SOLUTION 10-2

Let X represent the sentence, "You see a large thunderstorm in the distance." Let Y represent the sentence, "A thunderstorm is coming toward you." Consider this sentence:

$$X \Rightarrow Y$$

Now imagine that a thunderstorm is several miles away, moving from west-to-east, and you can see it. Therefore, variable X has truth value 1. But suppose you are located west of the storm, so it is moving away from you. Sentence Y has truth value 0. Therefore, the implication is false. If you see a thunderstorm, that doesn't necessarily mean it is coming toward you.

Table 10-4.   Truth table for logical implication.

X	Y	X $\Rightarrow$ Y
0	0	1
0	1	1
1	0	0
1	1	1

Table 10-5.   Truth table for logical equivalence (Boolean equality).

X	Y	X = Y
0	0	1
0	1	0
1	0	0
1	1	1

TABLE FOR LOGICAL EQUIVALENCE

If X and Y are logical variables, then X = Y has truth value 1 when both variables have value 1, or when both variables have value 0. If the truth values of X and Y are different, then X = Y has truth value 0. This is broken down fully in Table 10-5.

PROBLEM 10-3
Derive the truth table for logical equivalence based on the truth tables for conjunction and implication.

SOLUTION 10-3
Remember that X = Y means the same thing as (X ⇒ Y) × (Y ⇒ X). Based on this fact, you can build up X = Y in steps, as shown in Table 10-6 (proceeding from left to right). The four possible combinations of truth values for sentences X and Y are shown in the first (left-most) and second columns. The truth values for X ⇒ Y are shown in the third column, and the truth values for Y ⇒ X are shown in the fourth column. In order to get the truth values for the fifth or right-most column (X = Y), conjunction is applied to the truth values in the third and fourth columns.

Table 10-6.   Truth table for Problem 10-3.

X	Y	X ⇒ Y	Y ⇒ X	X = Y
0	0	1	1	1
0	1	1	0	0
1	0	0	1	0
1	1	1	1	1

Some Boolean Laws

Boolean operations obey certain *mathematical laws*, which means that they are inviolable. (If a single counterexample of a hypothesis is proven, then that hypothesis is not a mathematical law.) Here are some basic laws of Boolean algebra.

PRECEDENCE

When reading or constructing logical statements, the operations within parentheses are always performed first. If there are multilayered combinations of sentences (this is called *nesting of operations*), then you should first use ordinary parentheses, then square brackets [], and then curly brackets, also known as braces {}. Alternatively, you can use groups of plain parentheses inside each other. But if you do that, be sure you end up with the same number of left-hand parentheses and right-hand parentheses in the complete expression.

If there are no parentheses or brackets in an expression, instances of negation should be performed first. Then conjunctions (multiplication operations) should be done, then disjunctions (addition operations), then logical implications, and finally logical equivalences.

As an example of how precedence works, consider the following Boolean expression:

$$A \times -B + C \Rightarrow D$$

Using parentheses and brackets to clarify this according to the rules of precedence, we can write it like this:

$$\{[A \times (-B)] + C\} \Rightarrow D$$

Now consider a more complex Boolean expression, which is so messy that we run out of parenthesis and brackets if we use the "ordinary/square/curly" scheme:

$$A \times -B + C \Rightarrow D \times E = 0 + G$$

Using plain parentheses only, we can write it this way:

$$(((A \times (-B)) + C) \Rightarrow (D \times E)) = (0 + G)$$

When we count up the number of left-hand parentheses and the number of right-hand parentheses, we see that there are six left-hand ones and six right-hand ones. (If the number weren't the same, the expression would be flawed.)

CONTRADICTION

A *contradiction* always results in a false truth value (logic 0). This is one of the most interesting laws in mathematics, and has been used to prove important facts—and to construct satirical sentences. Symbolically, if X is any logical statement, we can write the rule like this:

$$(X \times -X) \Rightarrow 0$$

LAW OF DOUBLE NEGATION

The negation of a negation is equivalent to the original expression. That is, if X is any logical variable, then:

$$-(-X) = X$$

COMMUTATIVE LAWS

The conjunction of two variables always has the same value, regardless of the order in which the variables are expressed. If X and Y are logical variables, then X × Y is logically equivalent to Y × X:

$$X \times Y = Y \times X$$

The same property holds for logical disjunction:

$$X + Y = Y + X$$

These are called the *commutative law for conjunction* and the *commutative law for disjunction*, respectively. The variables can be commuted (interchanged in order) and it doesn't affect the truth value of the resulting sentence.

ASSOCIATIVE LAWS

When there are three variables combined by two conjunctions, it doesn't matter how the variables are grouped. Suppose you have a Boolean expression that can be symbolized as follows:

$$X \times Y \times Z$$

where X, Y, and Z represent the truth values of three constituent sentences. Then we can consider X × Y as a single variable and combine it with Z, or we can consider $Y \times Z$ as a single variable and combine it with X, and the results are logically equivalent:

$$(X \times Y) \times Z = X \times (Y \times Z)$$

The same law holds for logical disjunction:

$$(X + Y) + Z = X + (Y + Z)$$

These are called the *associative law for conjunction* and the *associative law for disjunction*, respectively.

We must be careful when applying associative laws. All the operations in the Boolean expression must be the same. If a Boolean expression contains a conjunction and a disjunction, we cannot change the grouping and expect to get the same truth value in all possible cases. For example, the following two Boolean expressions are not, in general, logically equivalent:

$$(X \times Y) + Z$$
$$X \times (Y + Z)$$

LAW OF IMPLICATION REVERSAL

When one sentence implies another, you can't reverse the sense of the implication and still expect the result to be valid. It is not always true that if $X \Rightarrow Y$, then $Y \Rightarrow X$. It can be true in certain cases, such as when $X = Y$. But there are plenty of cases where it isn't true.

If you negate both sentences and reverse the sense of the implication, however, the result is always valid. This is the *law of implication reversal*. It is also known as the *law of the contrapositive*. Expressed symbolically, suppose we are given two logical variables X and Y. Then the following always holds:

$$(X \Rightarrow Y) = (-Y \Rightarrow -X)$$

PROBLEM 10-4
Use words to illustrate an example of the above law in a real-world sense.

SOLUTION 10-4
Let V represent the sentence "Jane is a living vertebrate creature." Let B represent the sentence "Jane has a brain." Then $V \Rightarrow B$ reads, "If

Jane is a living vertebrate creature, then Jane has a brain." Applying the law of implication reversal, we can say that $-B \Rightarrow -V$. That translates to: "If Jane does not have a brain, then Jane is not a living vertebrate creature." These two sentences, although a bit strange, are logically equivalent.

DeMORGAN'S LAWS

If the conjunction of two sentences is negated as a whole, the result can be rewritten as the disjunction of the negations of the original two sentences. Expressed symbolically, if X and Y are two logical variables, then the following holds valid in all cases:

$$-(X \times Y) = (-X) + (-Y)$$

This is called *DeMorgan's law for conjunction*.

A similar rule holds for disjunction. If a disjunction of two sentences is negated as a whole, the resulting Boolean expression can be rewritten as the conjunction of the negations of the original two sentences. Symbolically:

$$-(X + Y) = (-X) \times (-Y)$$

This is called *DeMorgan's law for disjunction*.

DISTRIBUTIVE LAW

A specific relationship exists between conjunction and disjunction, known as the *distributive law*. It works like the *distributive principle* in arithmetic. That principle states that if *a* and *b* are any two numbers, then

$$a(b + c) = ab + ac$$

Think of logical conjunction as multiplication, and logical disjunction as addition. Then if X, Y, and Z are any three sentences, the following logical equivalence exists:

$$X \times (Y + Z) = (X \times Y) + (X \times Z)$$

This is called the *distributive law of conjunction with respect to disjunction*. Its resemblance to the arithmetic distributive principle can be used as a memory aid.

Quick Practice

Here are some practice problems that cover the material presented in this chapter. Solutions follow the problems.

PROBLEMS

1. Consider an operation called NAND, which consists of the AND operation acting on two variables, followed by negation of the result. Suppose this operation is indicated by a multiplication symbol with a circle around it ($\otimes$). Define this operation symbolically, and write down a truth table for it, as applicable to two variables.

2. Consider an operation called NOR, which consists of the inclusive OR operation acting on two variables, followed by negation of the result. Suppose this operation is indicated by an addition symbol with a circle around it ($\oplus$). Define this operation symbolically, and write down a truth table for it, as applicable to two variables.

3. Consider the exclusive OR operation (called XOR) that was defined earlier in this chapter. Suppose this operation is indicated by an addition symbol with a subtraction symbol underneath ($\pm$). Write down a truth table for it, as applicable to two variables.

4. Consider an operation called XNOR (or, alternatively, NXOR), which consists of the XOR operation acting on two variables, followed by negation of the result. Suppose this operation is indicated by the division symbol from arithmetic ($\div$). Define this operation symbolically, and write down a truth table for it, as applicable to two variables.

5. The truth values for the XNOR operation, as applied to two variables, are the same as those for a relation with which we are familiar. What relation is this? What is the difference (if any) between XNOR and this relation?

SOLUTIONS

1. Let X and Y be logical variables. If we let the NAND operation be symbolized $\otimes$, the following holds:

$$X \otimes Y = -(X \times Y)$$

Table 10-7. Truth table for Quick Practice Problem 1.

X	Y	X × Y	X ⊗ Y
0	0	0	1
0	1	0	1
1	0	0	1
1	1	1	0

Table 10-7 shows how the truth values for this operation are derived in terms of Boolean multiplication and negation.

2. Let X and Y be logical variables. If we let the NOR operation be symbolized ⊕, the following holds:

$$X \oplus Y = -(X + Y)$$

Table 10-8 shows how the truth values for this operation are derived in terms of Boolean addition and negation.

3. Let X and Y be logical variables. If we let the XOR operation be symbolized ±, then X ± Y = 1 when X ≠ Y, and X ± Y = 0 when X = Y. Table 10-9 shows the values of the XOR operation.

Table 10-8. Truth table for Quick Practice Problem 2.

X	Y	X + Y	X ⊕ Y
0	0	0	1
0	1	1	0
1	0	1	0
1	1	1	0

Table 10-9. Truth table for Quick Practice Problem 3.

X	Y	X ± Y
0	0	0
0	1	1
1	0	1
1	1	0

4. Let X and Y be logical variables. If we let the XOR operation be symbolized ±, and the XNOR operation be symbolized ÷, the following holds:

$$X \div Y = -(X \pm Y)$$

Table 10-10 shows how the truth values for this operation are derived in terms of XOR followed by negation.

5. The truth values for XNOR are the same as the truth values for logical equivalence. However, there is a difference between the two. While XNOR is an *operation*, which produces a resultant value on the basis of input variables, logical equivalence is a *relation*, which merely defines how the values of variables compare.

Table 10-10. Truth table for Quick Practice Problem 4.

X	Y	X ± Y	X ÷ Y
0	0	0	1
0	1	1	0
1	0	1	0
1	1	0	1

Quiz

This is an "open book" quiz. You may refer to the text in this chapter. A good score is 8 correct. Answers are in the back of the book.

1. The conjunction of five Boolean variables has value 0

    (a)   if and only if all the variables have value 0.
    (b)   if and only if at least one of the variables has value 0.
    (c)   if and only if all the variables have value 1.
    (d)   Forget it! A conjunction can't be defined for five variables.

2. The disjunction of seven variables has value 0

    (a)   if and only if all the variables have value 0.
    (b)   if and only if at least one of the variables has value 0.
    (c)   if and only if all the variables have value 1.
    (d)   Forget it! A disjunction can't be defined for seven variables.

3. In a logical statement, a double-shafted arrow pointing to the right means

    (a)   "and."
    (b)   "if."
    (c)   "if and only if."
    (d)   "logically implies."

4. How many possible combinations of truth values are there for a set of three variables, each of which can attain either the value 1 or the value 0?

    (a)   2
    (b)   4
    (c)   8
    (d)   16

5. Suppose you observe, "It is not sunny today, and it's not warm." Your friend says, "The statement that it's sunny or warm today is false." These two sentences are logically equivalent, and this constitutes a verbal example of

    (a)   one of DeMorgan's laws.
    (b)   the law of double negation.
    (c)   one of the commutative laws.
    (d)   the law of implication reversal.

6. Suppose I make the claim that a certain general statement is a law of logic. You demonstrate that my supposed law has at least one exception. This proves that

   (a) my statement is not a law of logic.
   (b) my statement violates the commutative law.
   (c) my statement violates the law of implication reversal.
   (d) a disjunction implies logical falsity.

7. Look at Table 10-11. What, if anything, is wrong with this truth table?

   (a) Not all possible combinations of truth values are shown for X, Y, and Z.
   (b) The entries in the far right-hand column are incorrect.
   (c) It is impossible to have a logical operation such as $(X + Y) \times Z$.
   (d) Nothing is wrong with Table 10-11.

8. What, if anything, can be done to make Table 10-11 show a valid derivation?

   (a) Nothing needs to be done. It is correct as it is.
   (b) In the top row, far-right column header, change the multiplication symbol ($\times$) to a double-shafted arrow pointing to the right ($\Rightarrow$).
   (c) In the far-left column, change every 0 to 1, and change every 1 to 0.
   (d) In the first three columns, change every 1 to 0, and change every 0 to 1.

Table 10-11.   Truth table for Quiz Questions 7 and 8

X	Y	Z	X + Y	$(X + Y) \times Z$
0	0	0	0	1
0	0	1	0	1
0	1	0	1	0
0	1	1	1	1
1	0	0	1	0
1	0	1	1	1
1	1	0	1	0
1	1	1	1	1

9. An atomic proposition in Boolean algebra

(a) can sometimes be split into smaller logical parts.
(b) can contain multiple Boolean operations.
(c) can always be split into smaller logical parts.
(d) can never be split into smaller logical parts.

10. Imagine that someone says to you, "If I am a human and I am not a human, then the moon is made of Swiss cheese." (Forget for a moment that this person has obviously lost contact with the real world.) This is a verbal illustration of the fact that

(a) implication can't be reversed.
(b) DeMorgan's laws don't always hold true.
(c) conjunction is not commutative.
(d) a contradiction implies logical falsity.

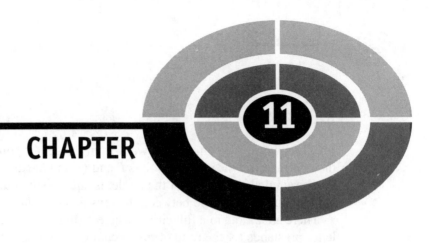

Trigonometric Functions

Trigonometry involves angles and their relationships to distances. All of these relationships arise from the characteristics of a circle, and can be defined on the basis of the graph of a circle in the Cartesian (rectangular coordinate) plane. The relationships can also be defined in terms of the relative dimensions of triangles.

The Unit Circle

Consider a circle in the Cartesian plane with the following equation:

$$x^2 + y^2 = 1$$

This is called the *unit circle* because its radius is 1 unit, and it is centered at the origin (0,0). This circle gives us a simple basis to define the common *trigonometric functions,* which are also called the *circular functions.*

RADIANS

Imagine two rays emanating outward from the center point of a circle. Suppose the rays intersect the circle at points P and Q, and the distance between P and Q, as measured along the arc of the circle, is equal to the radius of the circle. Then the measure of the angle between the rays is one *radian* (1 rad).

There are 2π rad in a full circle, where π (the lowercase, non-italicized Greek letter pi) stands for the ratio of the circumference of a circle to its diameter. The value of π is approximately 3.14159265359, often rounded off to 3.14159 or 3.14. The angle corresponding to a quarter-circle is $\pi/2$ rad; the angle corresponding to a half circle is π rad; the angle corresponding to three-quarters of a circle is $3\pi/2$ rad.

Mathematicians and physicists often use the radian when working with trigonometric functions, and the "rad" is sometimes left out when specifying angles in this form. If you see something like $\theta_1 = \pi/4$, you know the angle θ_1 is expressed in radians.

DEGREES, MINUTES, SECONDS

The *angular degree* (°), also called the *degree of arc,* is the unit of angular measure familiar to laypeople. One degree (1°) is the equivalent of 1/360 of a full circle. An angle of 90° represents a quarter circle, 180° represents a half circle, 270° represents three-quarters of a circle, and 360° represents a full circle. A *right angle* has a measure of 90°, an *acute angle* has a measure of more than 0° but less than 90°, and an *obtuse angle* has an angle more than 90° but less than 180°.

To denote the measures of tiny angles, or to precisely denote the measures of angles in general, smaller units are used. One *minute of arc* or *arc minute,* symbolized by an apostrophe or accent (′) or abbreviated as m or min, is (1/60)°. One *second of arc* or *arc second,* symbolized by a closing quotation mark (″) or abbreviated as s or sec, is (1/60)′ or (1/3600)°. An example of an angle in this notation is 30° 15′ 0″, which denotes 30 degrees, 15 minutes, 0 seconds.

Alternatively, fractions of a degree can be denoted in decimal form. You might see, for example, 30.25°. This is the same as 30° 15′ 0″. Decimal fractions of degrees are easier to work with than the /minute/second scheme when angles must be added and subtracted, or when using a conventional calculator to work out trigonometry problems.

PROBLEM 11-1

Suppose a textbook tells you that $\theta_1 = \pi/4$. What is the measure of θ_1 in degrees?

SOLUTION 11-1
There are 2π rad in a full circle of 360°. The value $\pi/4$ is equal to 1/8 of 2π. Therefore, the angle θ_1 is 1/8 of a full circle, or 45°.

PROBLEM 11-2
Suppose your town is listed in an almanac as being at 40° 20′ north latitude and 93° 48′ west longitude. What are these values in decimal form? Express your answers to two decimal places.

SOLUTION 11-2
There are 60 minutes of arc in an angular degree. For latitude, note that $20' = (20/60)° = 0.33°$; that means the latitude is 40.33° north. For longitude, note that $48' = (48/60)° = 0.80°$; that means the longitude is 93.80° west.

Primary Circular Functions

Consider a circle in rectangular coordinates with the following equation:

$$x^2 + y^2 = 1$$

This equation, as defined earlier in this chapter, represents the unit circle. Let θ be an angle whose apex is at the origin, and that is measured counterclockwise from the x axis, as shown in Fig. 11-1. Suppose this angle corresponds to a ray that intersects the unit circle at some point $P = (x_0, y_0)$. We can define three basic trigonometric functions, called circular functions, of the angle θ in a simple and elegant way.

THE SINE FUNCTION

Consider ray OP passing outward from the origin (point O) through a movable point P on the circle. Imagine this ray as pointing straight along the positive x axis, and then starting to rotate counterclockwise on its end point O, as if point O is a mechanical bearing. The point P, represented by coordinates (x_0, y_0), therefore revolves around point O, following the perimeter of the unit circle.

Imagine what happens to the value of y_0 (also known as the *ordinate* of point P) during one complete revolution of ray OP. The ordinate of P starts out at $y_0 = 0$, then increases until it reaches $y_0 = 1$ after P has gone 90° or $\pi/2$ rad

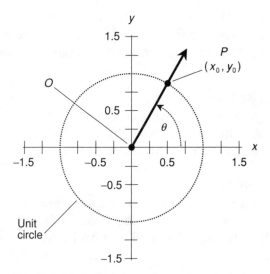

Fig. 11-1. The unit circle can be used as a basis for defining the trigonometric functions.

around the circle ($\theta = 90° = \pi/2$). After that, y_0 begins to decrease, getting back to $y_0 = 0$ when P has gone 180° or π rad around the circle ($\theta = 180° = \pi$). As P continues to revolve counterclockwise, y_0 keeps decreasing until, at $\theta = 270° = 3\pi/2$, the value of y_0 reaches its minimum of -1. After that, the value of y_0 rises again until, when P has gone completely around the circle, it returns to $y_0 = 0$ for $\theta = 360° = 2\pi$.

The value of y_0 is defined as the *sine* of the angle θ. The sine function is abbreviated sin, so we can state this simple equation:

$$\sin \theta = y_0$$

THE COSINE FUNCTION

Look again at Fig. 11-1. Imagine, once again, the ray *OP* from the origin outward through a movable point P on the circle. Think of the ray pointing directly along the positive x axis to start out with, and then rotating in a counterclockwise direction, as before.

Now imagine what happens to the value of x_0 (also called the *abscissa* of point P) during one complete revolution of ray OP. The abscissa of P starts out at $x_0 = 1$, then decreases until it reaches $x_0 = 0$ when $\theta = 90° = \pi/2$. After that, x_0 continues to decrease, getting down $x_0 = -1$ when $\theta = 180° = \pi$. As P continues counterclockwise around the circle, x_0 begins to increase again; at $\theta = 270° = 3\pi/2$, the value gets back up to $x_0 = 0$. After that, x_0 increases further until, when P has gone completely around the circle, it returns to $x_0 = 1$ for $\theta = 360° = 2\pi$.

The value of x_0 is defined as the *cosine* of the angle θ. The cosine function is abbreviated cos. So we can write this:

$$\cos \theta = x_0$$

THE TANGENT FUNCTION

Once again, refer to Fig. 11-1. The *tangent* (abbreviated tan) of an angle θ is defined using the same ray OP and the same point $P = (x_0, y_0)$ as is done with the sine and cosine functions. The definition is:

$$\tan \theta = y_0/x_0$$

Because we already know that $\sin \theta = y_0$ and $\cos \theta = x_0$, we can express the tangent function in terms of the sine and the cosine:

$$\tan \theta = \sin \theta / \cos \theta$$

This function "blows up" at certain values of θ. Whenever $x_0 = 0$, the denominator of either quotient above becomes zero. Division by zero is not defined, and that means the tangent function is not defined for any angle θ such that $\cos \theta = 0$. Such angles are all the odd multiples of $90°$ ($\pi/2$ rad).

PROBLEM 11-3
What is tangent of an exact $45°$ angle? Do not perform any calculations. You should be able to infer this without having to write down any numerals.

SOLUTION 11-3
Draw a diagram of a unit circle, such as the one in Fig. 11-1, and place ray OP such that it subtends an angle of exactly $45°$ with respect to the positive x axis. That angle is the angle of which we want to find the tangent. Note that the ray OP also subtends an angle of $45°$ with respect to the positive y axis, because the x and y axes are perpendicular (they

are oriented at 90° with respect to each other), and 45° is exactly half of 90°. Every point on the ray OP is equally distant from the positive x and y axes; this includes the point (x_0, y_0). It follows that $x_0 = y_0$, and neither of them is equal to zero. From this, we can conclude that $y_0/x_0 = 1$. According to the definition of the tangent function, therefore, the tangent of an exact 45° angle is precisely equal to 1.

Secondary Circular Functions

The three functions defined above form the cornerstone of trigonometry. However, three more circular functions exist. Their values represent the reciprocals of the values of the preceding three functions. To understand the definitions of these functions, look again at Fig. 11-1.

THE COSECANT FUNCTION

Imagine the ray OP, subtending an angle θ with respect to the x axis, and emanating out from the origin and intersecting the unit circle at the point $P = (x_0, y_0)$. The reciprocal of the ordinate, that is, $1/y_0$, is defined as the *cosecant* of the angle θ. The cosecant function is abbreviated csc, so we can state this simple equation:

$$\csc \theta = 1/y_0$$

This function is the reciprocal of the sine function. That is to say, for any angle θ, the following equation is always true as long as sin θ is not equal to 0:

$$\csc \theta = 1/\sin \theta$$

The cosecant function is not defined for 0° (0 rad), or for any multiple of 180° (π rad). This is because the sine of any such angle is equal to 0, which would mean that the cosecant would have to be equal to 1/0, which is an undefined quantity.

THE SECANT FUNCTION

Keeping the same rotating-ray notion in mind, consider $1/x_0$. This is defined as the secant of the angle θ. The secant function is abbreviated sec, so we can define it like this:

$$\sec \theta = 1/x_0$$

The secant of any angle is the reciprocal of the cosine of that angle. That is, as long as cos θ is not equal to 0, the following equation holds:

$$\sec \theta = 1/\cos \theta$$

The secant function is not defined for 90° ($\pi/2$ rad), or for any odd multiple thereof. This is because the cosine of any such angle is equal to 0, which would mean that the secant would have to be equal to 1/0, which is an undefined quantity.

THE COTANGENT FUNCTION

Now consider the value x_0/y_0. This is called the *cotangent* function, abbreviated cot. For any ray anchored at the origin and crossing the unit circle at an angle θ, the following holds:

$$\cot \theta = x_0/y_0$$

Because we already know that sin $\theta = y_0$ and cos $\theta = x_0$, we can express the cotangent function in terms of the cosine and the sine:

$$\cot \theta = \cos \theta / \sin \theta$$

The cotangent of an angle is also equal to the reciprocal of the tangent of that angle:

$$\cot \theta = 1/\tan \theta$$

Whenever $y_0 = 0$, the denominator of either quotient above becomes 0, and the cotangent function is not defined. This occurs at all integer multiples of 180° (π rad).

NONSTANDARD ANGLES

Once in awhile you will encounter an angle whose measure is negative, or whose measure is 360° (2π rad) or more. In trigonometry, any such angle can be reduced to something that is at least 0° (0 rad), but less than 360° (2π rad). Figure 11-1 shows why. Even if ray *OP* makes more than one revolution counterclockwise from the *x* axis, or if it turns clockwise, its *orientation* can always be defined by some counterclockwise angle of at least 0° (0 rad) but less than 360° (2π rad) relative to the positive *x* axis.

For the purposes of determining the values of trigonometric functions, any angle ϕ of the nonstandard sort, such as 450° or $-9\pi/4$ rad, can be reduced to an angle θ that is at least 0° (0 rad) but less than 360° (2π rad) by adding or subtracting some whole-number multiple of 360° (2π rad). Thus, for example:

$$\sin 450° = \sin (450 - 360)°$$
$$= \sin 90°$$
$$= 1$$

$$\cos (-9\pi /4) = \cos [-9\pi/4 + (2 \times 2\pi)]$$
$$= \cos (4\pi - 9\pi/4)$$
$$= \cos 7\pi/4$$
$$= 0.7071$$

PROBLEM 11-4

Use a portable scientific calculator, or the calculator program in a personal computer, to find the values of all six circular functions of 66°. Round your answers off to three decimal places. If your calculator does not have keys for the cosecant (csc), secant (sec), or cotangent (cot) functions, first find the sine (sin), cosine (cos), and tangent (tan) respectively, then find the reciprocal, and round off your answer to three decimal places as the final step.

SOLUTION 11-4

You should get the following results. Be sure your calculator is set to work with degrees, not radians.

$$\sin 66° = 0.914$$
$$\cos 66° = 0.407$$
$$\tan 66° = 2.246$$
$$\csc 66° = 1/\sin 66° = 1.095$$
$$\sec 66° = 1/\cos 66° = 2.459$$
$$\cot 66 = 1/\tan 66° = 0.445$$

The Right Triangle Model

In the previous section, we defined the six circular functions in terms of points on a circle. There is another way to define these functions: the *right-triangle model*.

TRIANGLE AND ANGLE NOTATION

In geometry, it is customary to denote triangles by writing an uppercase Greek letter delta (Δ) followed by the names of the three points representing the corners, or *vertices,* of the triangle. Angles are denoted by writing the symbol ∠ (which resembles an italicized, uppercase English letter *L* without serifs) followed by the names of three points that uniquely determine the angle. This scheme lets us specify the extent and position of the angle, and also the rotational sense in which it is expressed. For example, if there are three points *D*, *E*, and *F*, then ∠*FED* (read "angle *FED*") has the same measure as∠*DEF*, but in the opposite direction. The middle point, *E* in either case, is the *vertex* of the angle.

The rotational sense in which an angle is measured can be significant in physics, astronomy, and engineering, and also when working in coordinate systems. In the Cartesian plane, angles measured counterclockwise are considered positive by convention, while angles measured clockwise are considered negative. If we have ∠*FED* that measures 30° counterclockwise around a circle in Cartesian coordinates, then ∠*DEF*, which goes clockwise, measures −30°. The cosines of these two angles happen to be the same, but the sines differ.

RATIOS OF SIDES

Consider a right triangle defined by points *D*, *E*, and *F*, as shown in Fig. 11-2. Suppose that ∠*DFE* is a right angle (that is, it has a measure of 90°), so Δ*DEF* is a *right triangle.* Let *d* be the length of the line segment opposite point *D*. Let *e* be the length of line segment opposite point *E*. Let *f* be the length of line segment opposite point *F*. Let θ be ∠*FED*, the angle measured counterclockwise

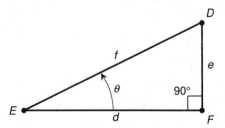

Fig. 11-2. The right-triangle model for defining trigonometric functions.

between line segments *EF* and *ED*. The six circular trigonometric functions can be defined as ratios between the lengths of the sides, as follows:

$$\sin \theta = e/f$$
$$\cos \theta = d/f$$
$$\tan \theta = e/d$$
$$\csc \theta = f/e$$
$$\sec \theta = f/d$$
$$\cot \theta = d/e$$

The longest side of a right triangle is always opposite the 90° angle, and is called the *hypotenuse*. In Fig. 11-2, this is the line segment *ED*, whose length is *f*. The other two sides are called *adjacent sides* because they are both adjacent to the right angle, and have a common vertex (in this case point *F*).

SUM OF ANGLE MEASURES

In any triangle, the sum of the measures of the interior angles is 180° (π rad). This holds true whether it is a right triangle or not, as long as all the angles are measured in the plane defined by the three vertices of the triangle.

RANGE OF ANGLES

In the right-triangle model, the values of the trigonometric functions are defined only for angles between (but not including) 0° and 90° (0 rad and $\pi/2$ rad). All angles outside this range are defined using the unit-circle model.

PROBLEM 11-5

Suppose there is a triangle whose sides are precisely 3, 4, and 5 units, respectively. What is the sine of the angle θ opposite the side that measures 3 units? Round the answer off to three significant figures.

SOLUTION 11-5

If we are to use the right-triangle model to solve this problem, we must first be certain that a triangle with sides having lengths of precisely 3, 4, and 5 units is a right triangle. We can test for this by seeing if the Pythagorean theorem applies. If this triangle is a right triangle, then the side measuring 5 units is the hypotenuse, and we should find that

$3^2 + 4^2 = 5^2$. Checking, we see that $3^2 = 9$ and $4^2 = 16$. Therefore, $3^2 + 4^2 = 9 + 16 = 25$, which is equal to 5^2. This is indeed a right triangle.

It helps to draw a picture here, after the fashion of Fig. 11-2. Put the angle θ, which we are analyzing, at lower left (corresponding to the vertex point E). Label the hypotenuse $f = 5$. Set $e = 3$ and $d = 4$. According to the formulas above, the sine of the angle in question is equal to e/f. In this case, that means $\sin \theta = 3/5 = 0.600$.

PROBLEM 11-6

What are the values of the other five circular functions for the angle θ as defined in Problem 11-5? Round the answers off to three significant figures.

SOLUTION 11-6

Simply plug numbers into the formulas given above, representing the ratios of the lengths of sides in the right triangle:

$$\cos \theta = d/f = 4/5 = 0.800$$
$$\tan \theta = e/d = 3/4 = 0.750$$
$$\csc \theta = f/e = 5/3 = 1.67$$
$$\sec \theta = f/d = 5/4 = 1.25$$
$$\cot \theta = d/e = 4/3 = 1.33$$

Trigonometric Identities

The following paragraphs depict common *trigonometric identities* for the circular functions. Unless otherwise specified, these formulas apply to angles θ and ϕ in the standard range, as follows:

$$0 \text{ rad} \le \theta < 2\pi \text{ rad}$$
$$0° \le \theta < 360°$$
$$0 \text{ rad} \le \phi < 2\pi \text{ rad}$$
$$0° \le \phi < 360°$$

Angles outside the standard range are converted to values within the standard range by adding or subtracting the appropriate multiple of 360° (2π rad). You will occasionally hear of an angle with negative measure, or with a measure of more than 360° (2π rad). Such an angle can always be converted to an equivalent angle with positive measure that is at least 0° (0 rad) but less than 360° (2π rad).

PYTHAGOREAN THEOREM FOR SINE AND COSINE

The sum of the squares of the sine and cosine of an angle is always equal to 1. The following formula holds:

$$\sin^2\theta + \cos^2\theta = 1$$

A NOTE ABOUT EXPONENTS

The expression $\sin^2\theta$ refers to the sine of the angle, squared (not the sine of the square of the angle). That is to say:

$$\sin^2\theta = (\sin\ \theta)^2$$

In general, the following notational definition applies, for any number x:

$$\sin^x\theta = (\sin\ \theta)^x$$

This rule also holds for all of the other trigonometric functions.

PYTHAGOREAN THEOREM FOR SECANT AND TANGENT

The following formulas apply for all angles except $\theta = 90°$ ($\pi/2$ rad) and $\theta = 270°$ ($3\pi/2$ rad):

$$\sec^2\theta - \tan^2\theta = 1$$
$$\tan^2\theta - \sec^2\theta = -1$$

PROBLEM 11-7
Use a drawing of the unit circle to help show why it is true that $\sin^2\theta + \cos^2\theta = 1$ for angles θ greater than $0°$ and less than $90°$. (Hint: a right triangle is involved.)

SOLUTION 11-7
Figure 11-3 is a drawing of the unit circle, with the angle θ defined counterclockwise between the x axis and a ray emanating from the origin. When the angle is greater than $0°$ but less than $90°$, a right triangle is formed, with a segment of the ray as the hypotenuse. The length of this segment is equal to the radius of the unit circle, or 1 unit. According to the Pythagorean theorem for right triangles, the square of the length of the hypotenuse is equal to the sum of the squares of

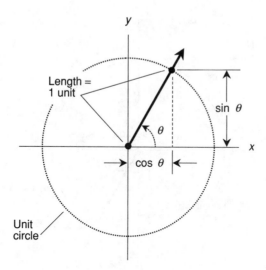

Fig. 11-3. Illustration for Problem 11-7.

the lengths of the other two sides. It is apparent from Fig. 11-3 that the lengths of these other two sides are sin θ and cos θ. Therefore:

$$(\sin\theta)^2 + (\cos\theta)^2 = 1^2$$

which is the same as saying that $\sin^2\theta + \cos^2\theta = 1$.

PROBLEM 11-8

Use another drawing of the unit circle to help show why it is true that $\sin^2\theta + \cos^2\theta = 1$ for angles θ greater than 270° and less than 360°. (Hint: this range of angles can be thought of as the range between, but not including, –90° and 0°).

SOLUTION 11-8

Figure 11-4 shows how this can be done. Draw a mirror image of Fig. 11-3, with the angle θ defined clockwise instead of counterclockwise. Again we have a right triangle; and this triangle, like all right triangles, has dimensions according to the Pythagorean theorem. Therefore:

$$(\sin\theta)^2 + (\cos\theta)^2 = 1^2$$

which is the same as saying that $\sin^2\theta + \cos^2\theta = 1$.

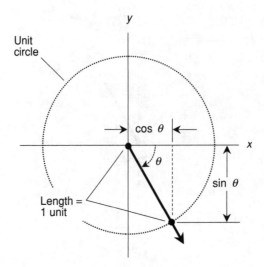

Fig. 11-4. Illustration for Problem 11-8.

SINE OF NEGATIVE ANGLE

The sine of the negative of an angle (an angle measured in the direction opposite to the normal direction) is equal to the negative of the sine of the angle. For all angles θ, the following formula holds:

$$\sin -\theta = -\sin \theta$$

COSINE OF NEGATIVE ANGLE

The cosine of the negative of an angle is equal to the cosine of the angle. For all angles θ, the following formula holds:

$$\cos -\theta = \cos \theta$$

TANGENT OF NEGATIVE ANGLE

The tangent of the negative of an angle is equal to the negative of the tangent of the angle. The following formula applies for all angles θ except 90° ($\pi/2$ rad) and 270° ($3\pi/2$ rad):

$$\tan -\theta = -\tan \theta$$

PROBLEM 11-9
Why doesn't the above formula work when $\theta = 90°$ ($\pi/2$ rad) or $\theta = 270°$ ($3\pi/2$ rad)?

SOLUTION 11-9
The value of the tangent function is not defined for those angles. Remember that the tangent of any angle is equal to the sine divided by the cosine. The cosine of 90° ($\pi/2$ rad) and the cosine of 270° ($3\pi/2$ rad) are both equal to 0. When a quotient has 0 in the denominator, that quotient is not defined. This is the reason for the restrictions on the angle measures in some of the formulas in the next few paragraphs.

COSECANT OF NEGATIVE ANGLE

The cosecant of the negative of an angle is equal to the negative of the cosecant of the angle. The following formula applies for all angles except $\theta = 0°$ (0 rad) and $\theta = 180°$ (π rad):

$$\csc -\theta = -\csc \theta$$

SECANT OF NEGATIVE ANGLE

The secant of the negative of an angle is equal to the secant of the angle. The following formula applies for all angles except $\theta = 90°$ ($\pi/2$ rad) and $\theta = 270°$ ($3\pi/2$ rad):

$$\sec -\theta = \sec \theta$$

COTANGENT OF NEGATIVE ANGLE

The cotangent of the negative of an angle is equal to the negative of the cotangent of the angle. The following formula applies for all angles θ except 0° (0 rad) and 180° (π rad):

$$\cot -\theta = -\cot \theta$$

SINE OF DOUBLE ANGLE

The sine of twice an angle θ is equal to twice the sine of the original angle times the cosine of the original angle:

$$\sin 2\theta = 2 \sin \theta \cos \theta$$

COSINE OF DOUBLE ANGLE

The cosine of twice an angle θ can be found using either of the following functions of the original angle:

$$\cos 2\theta = 1 - (2 \sin^2 \theta)$$
$$\cos 2\theta = (2 \cos^2 \theta) - 1$$

SINE OF ANGULAR SUM

The sine of the sum of two angles θ and ϕ can be found using this formula:

$$\sin (\theta + \phi) = \sin \theta \cos \phi + \cos \theta \sin \phi$$

COSINE OF ANGULAR SUM

The cosine of the sum of two angles θ and ϕ can be found using this formula:

$$\cos (\theta + \phi) = \cos \theta \cos \phi - \sin \theta \sin \phi$$

SINE OF ANGULAR DIFFERENCE

The sine of the difference between two angles θ and ϕ can be found using this formula:

$$\sin (\theta - \phi) = \sin \theta \cos \phi - \cos \theta \sin \phi$$

COSINE OF ANGULAR DIFFERENCE

The cosine of the difference between two angles θ and ϕ can be found using this formula:

$$\cos (\theta - \phi) = \cos \theta \cos \phi + \sin \theta \sin \phi$$

PROBLEM 11-10

Illustrate, using the unit circle model, examples of the following facts:
$$\sin -\theta = -\sin \theta$$
$$\cos -\theta = \cos \theta$$

SOLUTION 11-10

See Fig. 11-5. This shows an example for an angle θ of approximately 60° ($\pi/3$ rad). Note that the angle $-\theta$ is represented by rotation to the same extent as, but in the opposite direction from, the angle θ. Remember that positive angles are represented by counterclockwise rotation from the positive x axis, and negative angles are represented by clockwise rotation from the positive x axis. The ray from the origin for $-\theta$ is the reflection of the ray for θ with respect to the positive x axis. The above identities can be inferred geometrically from this diagram. The two rays intersect the circle at points whose y values (representing sines) are negatives of each other, and whose x values (representing cosines) are equal.

PROBLEM 11-11

Rewrite the expression $\sin (120° - \theta)$ as the sum of trigonometric functions of single angles. Express coefficients to three significant figures.

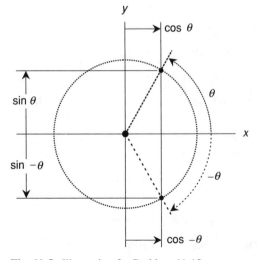

Fig. 11-5. Illustration for Problem 11-10.

SOLUTION 11-11

Use the formula for the sine of an angular difference, given above, substituting 120° for θ, as follows:

$$\sin (120° - \phi) = (\sin 120°)(\cos \phi) - (\cos 120°)(\sin \phi)$$
$$= 0.866 \cos \phi - (-0.500) \sin \phi$$
$$= 0.866 \cos \phi + 0.500 \sin \phi$$

In case you don't already know this definition, a *coefficient* is a number by which a variable or function is multiplied. In this situation, for example, the coefficients are 0.866 and 0.500.

PROBLEM 11-12

Illustrate, using the unit circle model, examples of the following facts:

$$\sin (180° - \theta) = \sin \theta$$
$$\cos (180° - \theta) = -\cos \theta$$

SOLUTION 11-12

See Fig. 11-6. This shows an example for an angle θ of approximately 30° ($\pi/6$ rad). The ray from the origin for 180° − θ is the reflection of the

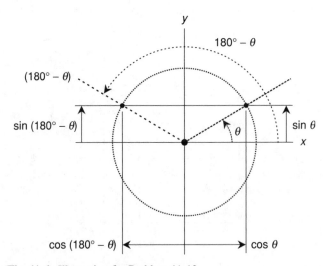

Fig. 11-6. Illustration for Problem 11-12.

ray for θ with respect to the positive y axis. The above identities can be inferred geometrically from this diagram. The two rays intersect the circle at points whose y values (representing sines) are equal, and whose x values (representing cosines) are negatives of each other.

Quick Practice

Here are some practice problems that cover the material presented in this chapter. Solutions follow the problems.

PROBLEMS

1. Sketch a graph of the equation $y = \sin x$ for values of x between -3π and 3π.

2. Sketch a graph of the equation $y = \cos x$ for values of x between -3π and 3π.

3. Sketch a graph of the equation $y = \tan x$ for values of x between -3π and 3π.

4. Sketch a graph of the equation $y = \csc x$ for values of x between -3π and 3π.

5. Sketch a graph of the equation $y = \sec x$ for values of x between -3π and 3π.

SOLUTIONS

1. See Fig. 11-7.

2. See Fig. 11-8.

3. See Fig. 11-9.

4. See Fig. 11-10.

5. See Fig. 11-11.

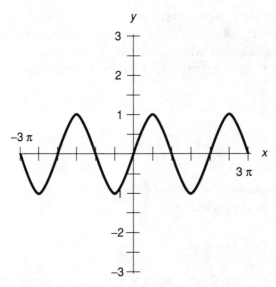

Fig. 11-7. Illustration for Quick Practice Problem and Solution 1.

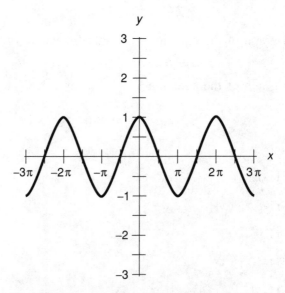

Fig. 11-8. Illustration for Quick Practice Problem and Solution 2.

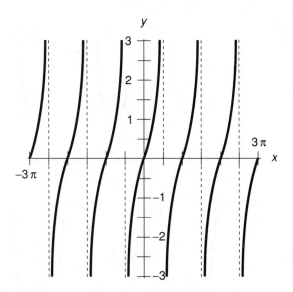

Fig. 11-9. IIllustration for Quick Practice Problem and Solution 3. The vertical dashed lines intersect the x axis at points where the function "blows up" and is undefined.

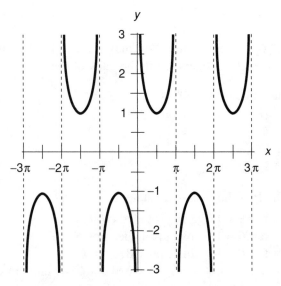

Fig. 11-10. Illustration for Quick Practice Problem and Solution 4. The vertical dashed lines intersect the x axis at points where the function "blows up" and is undefined.

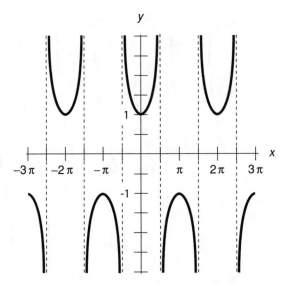

Fig. 11-11. Illustration for Quick Practice Problem and Solution 5. The vertical dashed lines intersect the x axis at points where the function "blows up" and is undefined.

Quiz

This is an "open book" quiz. You may refer to the text in this chapter. A good score is 8 correct. Answers are in the back of the book.

1. The value of tan 90° is
   (a) 0.
   (b) 1.
   (c) π.
   (d) not defined.

2. Which of the following statements is true?
   (a) tan θ = 1/cot θ, provided cot $\theta \neq 0$
   (b) tan θ = 1 − cos θ, provided cos $\theta \neq 0$
   (c) tan θ = 1 + sin θ, provided sin $\theta \neq 0$
   (d) tan θ + cot θ = 0, provided cot $\theta \neq 0$ and tan $\theta \neq 0$

3. With regard to the circular functions, an angle of 5π rad can be considered equivalent to an angle of

(a) 0°.
(b) 90°.
(c) 180°.
(d) 270°.

4. Suppose the tangent of a certain angle is –1.0000, and its cosine is –0.7071, approximated to four decimal places. The sine of this angle, approximated to four decimal places, is

(a) 1.0000.
(b) 0.7071.
(c) –0.7071.
(d) 0.0000.

5. What is the approximate measure of the angle described in Question 4?

(a) 0°
(b) 90°
(c) 180°
(d) None of the above

6. Refer to Fig. 11-12. The tangent of ∠ABC is equal to

(a) the length of line segment AC divided by the length of line segment AB.
(b) the length of line segment AD divided by the length of line segment BD.
(c) the length of line segment AD divided by the length of line segment AB.
(d) no ratio of lengths that can be shown here.

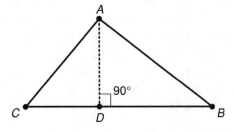

Fig. 11-12. Illustration for Quiz Questions 6 through 10.

7. Refer to Fig. 11-12. Suppose we know that the measure of ∠*BCA* is 50° and the length of line segment *AD* is 5.3 units. What is the length of line segment *AC*? Express the answer to one decimal place (that is, the nearest tenth of a unit).

   (a) 6.9 units

   (b) 8.2 units

   (c) 6.3 units

   (d) More information is needed to determine the answer.

8. Refer again to Fig. 11-12. Suppose we know that the measure of ∠*BCA* is 50° and the length of line segment *AD* is 5.3 units. What is the length of line segment *AB*? Express the answer to one decimal place (that is, the nearest tenth of a unit).

   (a) 6.9 units

   (b) 8.2 units

   (c) 6.3 units

   (d) More information is needed to determine the answer.

9. Refer again to Fig. 11-12. Suppose we know that line segment *AD* is exactly 2/3 as long as line segment *AB*. What is the measure of ∠*DAB*? Express the answer to the nearest tenth of a degree. Use a calculator if necessary.

   (a) 33.7°

   (b) 41.8°

   (c) 48.2°

   (d) 56.3°

10. Suppose, in reference to Fig. 11-12, we are told that the measure of ∠*BCA* is 50° and the measure of ∠*ABC* is 38°. We think that the person who says this is mistaken because

   (a) it would imply that Δ*ABC* is a right triangle, which is impossible.

   (b) it would imply that the measure of ∠*CAD* is something other than 40°, but it must be 40° because the sum of the measures of the interior angles of any triangle is 180°.

   (c) the measure of ∠*ABC* is 40° because Δ*ABC* is an isosceles triangle.

   (d) of a rush to judgment! It is entirely possible that the measure of ∠*BCA* is 50° and the measure of ∠*ABC* is 38°.

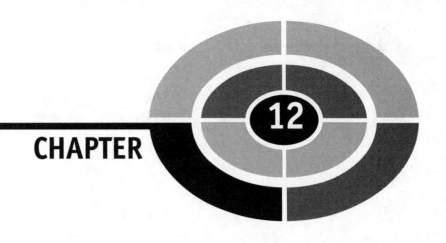

Vectors in Two and Three Dimensions

A *vector* is a quantity with two independent properties: *magnitude* and *direction*. In equations, vectors are denoted as letters in bold type, with lines over them, or with arrows over them. (In this book, we'll use lowercase boldface letters.) Geometrically, a vector is portrayed as a line segment, with length indicating the magnitude and an arrow indicating the direction with respect to a *reference axis*.

Vectors in the Cartesian Plane

Consider two arbitrary vectors, **a** and **b**, extending in the *xy*-plane from the origin (0,0) to points (x_a, y_a) and (x_b, y_b), as shown in Fig. 12-1. Here are some important definitions and properties of these vectors.

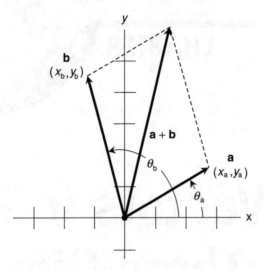

Fig. 12-1.   Two vectors, **a** and **b**, in the xy-plane.

MAGNITUDE IN xy-PLANE

The magnitude of **a**, written $|\mathbf{a}|$ or a, is given by the following formula:

$$|\mathbf{a}| = (x_a^2 + y_a^2)^{1/2}$$

DIRECTION IN xy-PLANE

When $x_a > 0$, the direction of **a** (written dir **a**) is the angle θ_a that the vector **a** subtends, expressed in a counterclockwise sense, from the $+x$ axis, as follows:

$$\text{dir } \mathbf{a} = \theta_a$$
$$= \arctan (y_a/x_a)$$
$$= \tan^{-1} (y_a/x_a)$$

When $x_a < 0$, the formula for angles in degrees is:

$$\text{dir } \mathbf{a} = \theta_a$$
$$= 180° + \arctan (y_a/x_a)$$
$$= 180° + \tan^{-1} (y_a/x_a)$$

When $x_a < 0$, the formula for angles in radians is:

$$\text{dir } \mathbf{a} = \theta_a$$
$$= \pi + \arctan (y_a/x_a)$$
$$= \pi + \tan^{-1} (y_a/x_a)$$

If $x_a = 0$, then dir $\mathbf{a} = 90°$ (or $\pi/2$ rad) for $y_a > 0$, and dir $\mathbf{a} = 270°$ (or $3\pi/2$ rad) for $y_a < 0$. If $x_a = 0$ and $y_a = 0$, then dir $\mathbf{a}$ is undefined.

SUM IN *xy*-PLANE

The sum of two vectors $\mathbf{a}$ and $\mathbf{b}$ in the Cartesian plane is given by the following formula:

$$\mathbf{a} + \mathbf{b} = [(x_a + x_b),(y_a + y_b)]$$

This sum can be found geometrically by constructing a parallelogram with $\mathbf{a}$ and $\mathbf{b}$ as adjacent sides. The sum vector, $\mathbf{a} + \mathbf{b}$, corresponds to the diagonal of this parallelogram.

DOT PRODUCT IN *xy*-PLANE

The *dot product* $\mathbf{a} \bullet \mathbf{b}$ of two vectors $\mathbf{a}$ and $\mathbf{b}$ in the Cartesian plane is a real number given by the formula:

$$\mathbf{a} \bullet \mathbf{b} = x_a x_b + y_a y_b$$

The dot product is also known as the *scalar product* because it is a scalar quantity, having magnitude but not direction.

CROSS PRODUCT IS PERPENDICULAR TO *xy*-PLANE

The *cross product* $\mathbf{a} \times \mathbf{b}$ of two vectors $\mathbf{a}$ and $\mathbf{b}$ is a vector oriented at a right angle to the plane containing $\mathbf{a}$ and $\mathbf{b}$. Let θ_{ab} be the angle between vectors $\mathbf{a}$ and $\mathbf{b}$, as expressed in the counterclockwise sense in the plane containing them both, as you look straight down on that plane. That is, $\theta_{ab} = (\theta_b - \theta_a)$. The magnitude of $\mathbf{a} \times \mathbf{b}$ is then given by the following formula:

$$|\mathbf{a} \times \mathbf{b}| = |\mathbf{a}|\,|\mathbf{b}|\sin\theta_{ab}$$

If the direction angle θ_b is greater than the direction angle θ_a (as shown in Fig. 12-1), then $\mathbf{a} \times \mathbf{b}$ points toward you as you look straight down on the xy-plane. If $\theta_b < \theta_a$, then $\mathbf{a} \times \mathbf{b}$ points away from you as you look straight down on the xy-plane.

The vector $\mathbf{b} \times \mathbf{a}$ has the same magnitude as the vector $\mathbf{a} \times \mathbf{b}$, but points in exactly the opposite direction. The cross product is often called the *vector product* because it is a vector quantity, having both magnitude and direction.

RIGHT-HAND RULE

Imagine two vectors $\mathbf{a}$ and $\mathbf{b}$. Suppose the angle between them, when their *originating points* (also called *back-end points*) coincide, is θ_{ab}. Curl the fingers of your right hand and point your thumb out so it is perpendicular to the plane containing your curled-up index finger. Then position your right hand so your fingers curl in the rotational sense of θ_{ab} (counterclockwise), while the plane containing your curled-up index finger coincides with the plane containing $\mathbf{a}$ and $\mathbf{b}$. When you do this, your thumb points in the general direction of $\mathbf{a} \times \mathbf{b}$, and in the opposite general direction from $\mathbf{b} \times \mathbf{a}$.

The vectors $\mathbf{a} \times \mathbf{b}$ and $\mathbf{b} \times \mathbf{a}$ are always oriented exactly perpendicular (also called *normal* or *orthogonal*) to the plane defined by $\mathbf{a}$ and $\mathbf{b}$, when $\mathbf{a}$ and $\mathbf{b}$ are positioned so that their originating points coincide.

PROBLEM 12-1
Consider two vectors in the Cartesian plane, $\mathbf{a} = (4,0)$ and $\mathbf{b} = (3,4)$. What is their sum vector, $\mathbf{a} + \mathbf{b}$? What is their dot product, $\mathbf{a} \bullet \mathbf{b}$? Assume that the values given are exact.

SOLUTION 12-1
Let $x_a = 4$, $x_b = 3$, $y_a = 0$, and $y_b = 4$. The sum vector is found by adding the components:

$$\mathbf{a} + \mathbf{b} = [(4 + 3),(0 + 4)]$$
$$= (7,4)$$

The dot product can be found by plugging the numbers into the formula for calculating dot products:

$$\mathbf{a} \bullet \mathbf{b} = (4 \times 3) + (0 \times 4)$$
$$= 12 + 0$$
$$= 12$$

PROBLEM 12-2

What is $|\mathbf{a} \times \mathbf{b}|$ for the vectors $\mathbf{a}$ and $\mathbf{b}$ defined in Problem 12-1? In what direction does $\mathbf{a} \times \mathbf{b}$ point? Assume the values given are exact.

SOLUTION 12-2

In order to find $|\mathbf{a} \times \mathbf{b}|$, we must know the lengths of both vectors, as well as the sine of the angle between them. If it helps you, feel free to draw a diagram to accompany this solution. To find $|\mathbf{a}|$ and $|\mathbf{b}|$, proceed as follows:

$$|\mathbf{a}| = (4^2 + 0^2)^{1/2}$$
$$= 16^{1/2}$$
$$= 4$$

$$|\mathbf{b}| = (3^2 + 4^2)^{1/2}$$
$$= (9 + 16)^{1/2}$$
$$= 25^{1/2}$$
$$= 5$$

In order to determine the sine of the angle between the two vectors, consider a right triangle formed by the positive x axis (in which $\mathbf{a}$ is contained), the vertical line $x = 3$, and the vector $\mathbf{b}$. The sine of the angle θ_{ab} between the vectors at the origin is equal to the height of this triangle, or y_b, divided by the length of the hypotenuse, or $|\mathbf{b}|$. Therefore:

$$\sin \theta_{ab} = y_b / |\mathbf{b}|$$
$$= 4/5$$

To find the magnitude of the cross product vector, use the formula for that purpose:

$$|\mathbf{a} \times \mathbf{b}| = |\mathbf{a}| \, |\mathbf{b}| \sin \theta_{ab}$$
$$= 4 \times 5 \times 4/5$$
$$= 16$$

If you look straight down on the Cartesian plane containing the vectors, then $\mathbf{a} \times \mathbf{b}$ is perpendicular to that plane, and points directly toward you.

PROBLEM 12-3

What is the magnitude of the cross product vector, $|\mathbf{b} \times \mathbf{a}|$, of the two vectors **a** and **b** defined in Problem 12-1? In what direction does $\mathbf{b} \times \mathbf{a}$ point?

SOLUTION 12-3

The value of $|\mathbf{b} \times \mathbf{a}|$ is the same as the value of $|\mathbf{a} \times \mathbf{b}|$. This is easy to prove, as follows:

$$|\mathbf{b} \times \mathbf{a}| = |\mathbf{b}|\,|\mathbf{a}|\,\sin\theta_{ab}$$
$$= |\mathbf{a}|\,|\mathbf{b}|\,\sin\theta_{ab}$$
$$= |\mathbf{a} \times \mathbf{b}|$$
$$= 16$$

The direction of $\mathbf{b} \times \mathbf{a}$ is precisely opposite to the direction of $\mathbf{a} \times \mathbf{b}$. If you look straight down on the Cartesian plane containing the vectors, then $\mathbf{b} \times \mathbf{a}$ is perpendicular to that plane, and points directly away from you.

Vectors in the Polar Plane

In the polar coordinate plane, vectors **a** and **b** can be denoted as rays from the origin (0,0) to points (θ_a, r_a) and (θ_b, r_b), as shown in Fig. 12-2. Angles in this illustration are portrayed in radians.

MAGNITUDE AND DIRECTION

The magnitude and direction of the vector $\mathbf{a} = (\theta_a, r_a)$ the polar coordinate plane are defined as follows:

$$|\mathbf{a}| = r_a$$

$$\text{dir } \mathbf{a} = \theta_a$$

By convention, these restrictions hold:

$$r_a \geq 0$$

$$0° \leq \theta_a < 360°$$

$$0 \text{ rad} \leq \theta_a < 2\pi \text{ rad}$$

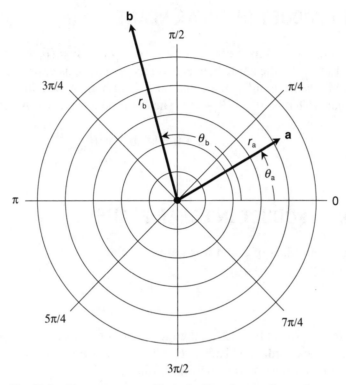

Fig. 12-2.   Two vectors, **a** and **b**, in the 2D polar coordinate plane.

SUM IN POLAR PLANE

The sum of two vectors **a** and **b**, when they are given in polar form, can be found by converting both vectors into rectangular (*xy*-plane) coordinates, adding the vectors according to the formula for vector addition in the *xy*-plane, and then converting the sum vector back to polar coordinates. To convert vector **a** from polar to rectangular coordinates, these formulas apply:

$$x_a = r_a \cos \theta_a$$

$$y_a = r_a \sin \theta_a$$

To convert vector **a** from rectangular coordinates to polar coordinates, use the formulas given earlier in this chapter for the direction and magnitude of a vector in the *xy*-plane. The value of dir **a** thus derived is equal to θ_a, and the value of |**a**| is equal to r_a.

DOT PRODUCT IN POLAR PLANE

In the polar coordinate plane, let $\mathbf{a} = (\theta_a, r_a)$ and $\mathbf{b} = (\theta_b, r_b)$, as shown in Fig. 12-2. Let θ_{ab} be the angle between vectors $\mathbf{a}$ and $\mathbf{b}$, as expressed in the counter-clockwise sense in the plane containing them both, as you look straight down on that plane. That is, $\theta_{ab} = (\theta_b - \theta_a)$. The dot product of $\mathbf{a}$ and $\mathbf{b}$ can be found using this formula:

$$\mathbf{a} \cdot \mathbf{b} = |\mathbf{a}|\,|\mathbf{b}| \cos \theta_{ab}$$
$$= r_a r_b \cos \theta_{ab}$$

CROSS PRODUCT IN POLAR PLANE

The cross product of $\mathbf{a}$ and $\mathbf{b}$ is perpendicular to the polar plane. Its magnitude is given by:

$$|\mathbf{a} \times \mathbf{b}| = |\mathbf{a}|\,|\mathbf{b}| \sin \theta_{ab}$$
$$= r_a r_b \sin \theta_{ab}$$

If $\theta_b > \theta_a$ (as in Fig. 12-2), then $\mathbf{a} \times \mathbf{b}$ points toward you as you look straight down on the coordinate plane. If $\theta_b < \theta_a$, then $\mathbf{a} \times \mathbf{b}$ points away from you as you look straight down on the coordinate plane.

PROBLEM 12-4

Consider two vectors $\mathbf{a}$ and $\mathbf{b}$ in the polar plane, with coordinates (θ_a, r_a) and (θ_b, r_b) as follows, with angles specified in degrees:

$$\mathbf{a} = (30°, 3)$$

$$\mathbf{b} = (150°, 2)$$

What is the dot product, $\mathbf{a} \cdot \mathbf{b}$? Assume the values given are exact.

SOLUTION 12-4

The calculation of this is straightforward, using the formula for dot product in polar coordinates. Note that $\theta_{ab} = \theta_b - \theta_a = 150° - 30° = 120°$. Therefore:

$$\mathbf{a} \cdot \mathbf{b} = r_a r_b \cos \theta_{ab}$$
$$= 3 \times 2 \times \cos 120°$$
$$= 3 \times 2 \times (-0.5)$$
$$= -3$$

PROBLEM 12-5

Consider two vectors **a** and **b** in the polar plane, with coordinates (θ_a, r_a) and (θ_b, r_b) as follows, with angles specified in radians:

$$\mathbf{a} = (\pi, 6)$$

$$\mathbf{b} = (\pi/4, 7)$$

What is the cross product, $\mathbf{a} \times \mathbf{b}$? Assume the values given are exact. Consider $\pi = 3.14159$, and determine the answer to three significant figures.

SOLUTION 12-5

This problem is a little tricky, because the quantity θ_{ab}, which is the quantity $(\theta_b - \theta_a)$, is negative:

$$\theta_b - \theta_a = \pi/4 - \pi$$
$$= -3\pi/4$$

In this case, the magnitude of the cross product vector is determined as follows:

$$|\mathbf{a} \times \mathbf{b}| = r_a r_b \sin \theta_{ab}$$
$$= 6 \times 7 \times \sin (-3\pi/4)$$
$$= 6 \times 7 \times -0.7071$$
$$= -29.7$$

In order to determine the direction of $\mathbf{a} \times \mathbf{b}$, note that if you look straight down on the polar plane, the angular movement (that is, the rotational sense) as you go from vector **a** to vector **b** is clockwise rather than counterclockwise. The calculation yields "−29.7 units toward you" because the sense of angular rotation is reversed from normal (clockwise rather than counterclockwise). By convention, vector magnitudes are non-negative. The correct way to define this vector is "29.7 units away from you, perpendicular to the polar plane."

Vectors in Cartesian 3-Space

In *xyz*-space, vectors **a** and **b** can be denoted as rays from the origin (0,0,0) to points (x_a, y_a, z_a) and (x_b, y_b, z_b), as shown in Fig. 12-3. The sum vector, $\mathbf{a} + \mathbf{b}$, is also shown. This is a perspective drawing. All three vectors would point generally out of the page toward you in a true 3D rendition.

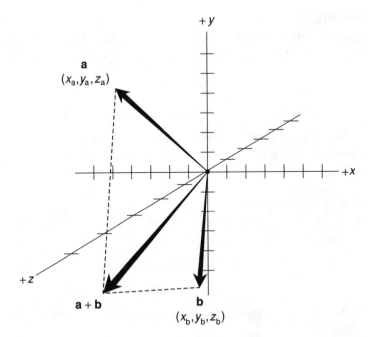

Fig. 12-3.   Two vectors, **a** and **b**, xyz-space. Their sum vector, **a** + **b**, is also shown. (The vector-addition parallelogram is distorted by the perspective.)

CORRECT ORIENTATION OF AXES

When xyz-space is portrayed, the $+z$ axis should point toward you when the $+x$ axis points to your right and the $+y$ axis points upward. Each of the three coordinate axes is perpendicular to the other two at the origin $(0,0,0)$. In properly constructed xyz-space, the cross product $(1,0,0) \times (0,1,0)$ is equal to the vector $(0,0,1)$. It's important that the orientations of the $+x$, $+y$, and $+z$ axes be in the correct relative sense. Otherwise, incorrect results will be obtained when determining dot products and cross products.

SUM IN xyz-SPACE

In Cartesian 3-space, the sum of two vectors $\mathbf{a} = (x_a, y_a, z_a)$ and $\mathbf{b} = (x_b, y_b, z_b)$ is given by the following formula:

$$\mathbf{a} + \mathbf{b} = [(x_a + x_b), (y_a + y_b), (z_a + z_b)]$$

This sum can be found geometrically by constructing a parallelogram with **a** and **b** as adjacent sides. The sum **a** + **b** is the diagonal of the parallelogram. An example is shown in the perspective drawing of Fig. 12-3.

MAGNITUDE IN *xyz*-SPACE

The magnitude of vector **a** = (x_a, y_a, z_a) in Cartesian 3-space, written |**a**| or a, is given by:

$$|\mathbf{a}| = (x_a^2 + y_a^2 + z_a^2)^{1/2}$$

DIRECTION IN *xyz*-SPACE

The direction of a vector **a** in Cartesian 3-space is denoted by specifying the angles θ_x, θ_y, and θ_z that **a** subtends relative to the +x, +y, and +z axes respectively, as shown in Fig. 12-4. These angles, expressed as an ordered triple $(\theta_x, \theta_y, \theta_z)$, are called the *direction angles* of the vector **a**. By convention, the rotational sense in space is irrelevant when specifying direction angles. Thus, direction angles are never negative, and they are never larger than 180° (π rad). With this scheme, all possible vector orientations can be uniquely defined in Cartesian 3-space.

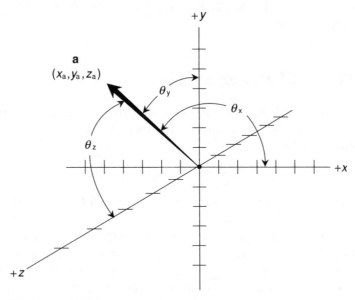

Fig. 12-4.   Direction angles of a vector in *xyz*-space.

Sometimes the cosines of the direction angles are specified, rather than the measures of the angles, to define the orientation of a vector in xyz-space. These are denoted using the lowercase Greek letters alpha (α), beta (β), and gamma (γ), and are known as the *direction cosines* of **a**:

$$\text{dir } \mathbf{a} = (\alpha, \beta, \gamma)$$

where $\alpha = \cos \theta_x$, $\beta = \cos \theta_y$, and $\gamma = \cos \theta_z$. Direction cosine values can range anywhere between, and including, -1 and 1.

DOT PRODUCT IN *xyz*-SPACE

The dot product $\mathbf{a} \bullet \mathbf{b}$ of two vectors $\mathbf{a} = (x_a, y_a, z_a)$ and $\mathbf{b} = (x_b, y_b, z_b)$ in xyz-space is a real number:

$$\mathbf{a} \bullet \mathbf{b} = x_a x_b + y_a y_b + z_a z_b$$

Alternatively, the dot product $\mathbf{a} \bullet \mathbf{b}$ of two vectors **a** and **b** in Cartesian 3-space is a real-number quantity given by this formula:

$$|\mathbf{a} \times \mathbf{b}| = |\mathbf{a}|\, |\mathbf{b}| \cos \theta_{ab}$$

where θ_{ab} is the angle between **a** and **b** as determined in the plane P containing them both, expressed in the rotational sense starting at **a** and ending up at **b**, as seen by an external observer.

CROSS PRODUCT IN *xyz*-SPACE

The cross product $\mathbf{a} \times \mathbf{b}$ of two vectors **a** and **b** in Cartesian 3-space is a vector perpendicular to the plane P containing both **a** and **b**, and whose magnitude is given by this formula:

$$|\mathbf{a} \times \mathbf{b}| = |\mathbf{a}|\, |\mathbf{b}| \sin \theta_{ab}$$

where θ_{ab} is the angle between **a** and **b** as determined in P, expressed in the rotational sense starting at **a** and ending up at **b**, as seen by an external observer.

The vector $\mathbf{a} \times \mathbf{b}$ is always perpendicular to the plane P that contains both **a** and **b**. If **a** and **b** are observed from some point on a line perpendicular to P and intersecting P at the origin, and θ_{ab} is expressed in a counterclockwise rotational sense from **a** to **b**, then $\mathbf{a} \times \mathbf{b}$ points toward the observer. If **a** and **b** are observed from some point on a line perpendicular to P and intersecting P at the origin, and

θ_{ab} is expressed in a clockwise rotational sense from **a** to **b**, then **a** × **b** points away from the observer.

PROBLEM 12-6

Consider the vector **a** = (2.000,3.000,4.000) in Cartesian 3-space. What is the magnitude of this vector? Express the answer to the appropriate number of significant figures.

SOLUTION 12-6

Let x_a = 2.000, y_a = 3.000, and z_a = 4.000. Plug these values into the formula for the magnitude of a vector in xyz-space, and calculate as follows:

$$|a| = (x_a^2 + y_a^2 + z_a^2)^{1/2}$$
$$= (2.000^2 + 3.000^2 + 4.000^2)^{1/2}$$
$$= (4.000 + 9.000 + 16.00)^{1/2}$$
$$= 29.00^{1/2}$$
$$= 5.385$$

PROBLEM 12-7

Consider two vectors **a** = (2,3,4) and **b** = (−1,5,0). What is **a** • **b**? Assume the values are exact. Express the answer to four significant figures.

SOLUTION 12-7

Let x_a = 2, y_a = 3, z_a = 4, x_b = −1, y_b = 5, and z_b = 0. Plug these values into the formula for the dot product of two vectors in xyz-space, and calculate as follows:

$$\mathbf{a} \cdot \mathbf{b} = x_a x_b + y_a y_b + z_a z_b$$
$$= (2 \times -1) + (3 \times 5) + (4 \times 0)$$
$$= -2 + 15 + 0$$
$$= 13.00$$

PROBLEM 12-8

Consider two vectors **a** and **b** in Cartesian 3-space. Suppose they both have magnitude 2.000, but their directions differ by 20°. What is |**a** × **b**|? Express the answer to the appropriate number of significant figures.

SOLUTION 12-8

Let $\theta_{ab} = 20°$. Plug the numbers into the formula for the magnitude of the cross product of two vectors in 3D, and calculate as follows:

$$|\mathbf{a} \times \mathbf{b}| = |\mathbf{a}| \, |\mathbf{b}| \sin \theta_{ab}$$
$$= 2.000 \times 2.000 \times \sin 20°$$
$$= 4.000 \times 0.34202014$$
$$= 1.368$$

PROBLEM 12-9

Show that the cross product of any two vectors that point in the same direction, regardless of their magnitudes, is the zero vector.

SOLUTION 12-9

When two vectors $\mathbf{a}$ and $\mathbf{b}$ point in the same direction, the angle θ_{ab} between them is $0°$. In such a situation, $|\mathbf{a} \times \mathbf{b}|$ is determined as follows:

$$|\mathbf{a} \times \mathbf{b}| = |\mathbf{a}| \, |\mathbf{b}| \sin \theta_{ab}$$
$$= |\mathbf{a}| \, |\mathbf{b}| \sin 0°$$
$$= |\mathbf{a}| \times |\mathbf{b}| \times 0$$
$$= 0$$

Therefore, $\mathbf{a} \times \mathbf{b} = \mathbf{0}$ when the vectors $\mathbf{a}$ and $\mathbf{b}$ have the same direction. (The zero vector is denoted by a boldface numeral $\mathbf{0}$.)

You might wish to show, as an additional exercise, that the cross product of any two vectors that point in opposite directions is equal to the zero vector. With this knowledge, we can say that any vector $\mathbf{a}$ has the following three properties:

$$\mathbf{a} \times \mathbf{a} = \mathbf{0}$$
$$-\mathbf{a} \times \mathbf{a} = \mathbf{0}$$
$$\mathbf{a} \times -\mathbf{a} = \mathbf{0}$$

Standard Form of a Vector

In any coordinate system, vectors can be geometrically expressed as finite-length rays with originating points that coincide with the coordinate origin. This is the *standard form of a vector*. In standard form, a vector can be depicted as a set of coordinates such as $(x,y,z) = (3,-5,5)$ or $(\theta,r) = (\pi/4,10)$.

EQUIVALENT AND IDENTICAL VECTORS

Two vectors **a** and **b** are *equivalent* if and only if they both have the same magnitude and the same direction. Two vectors **a** and **b** are *identical* if and only if they both have the same magnitude, the same direction, and the same originating point. The equality symbol (=) can be used to indicate either situation, as long as it's made clear whether the symbol refers to vectors that are truly identical, or to vectors that are merely equivalent.

In physics and engineering, magnitude and direction are sometimes the only important factors in the expression of a vector quantity. The actual originating point is not relevant in these cases. Then two equivalent vectors **a** and **b** can be considered equal, and this is written **a** = **b**. Examples include vectors representing the velocity or acceleration of a moving vehicle, when the location of the vehicle is not specified.

In pure mathematics, and in general science when the actual originating point is important, two vectors in the *xy*-plane or in *xyz*-space can be equivalent, but not identical. Then we must be careful about calling two vectors "equal." In this interpretation, every vector **a** has an infinite number of equivalent vectors **a′**, where the magnitudes and directions are all the same, but the originating points are all different.

In the discussions and examples that follow, all vectors are assumed to be in standard form, so their originating points are always at the origin of the coordinate system, unless otherwise specified. This will eliminate any possible confusion about what is meant when it is said that two vectors are "equal." When vectors are expressed in standard form, equivalent vectors are always identical. Then when we write an expression that claims two vectors are "equal" (for example **a** = **b**), it means that they are identical.

IN THE *xy*-PLANE

In two-dimensional Cartesian coordinates, let P_1 be the originating point of some vector **a′**, and let P_2 be the *terminating point* of that same vector **a′**, defined as follows:

$$P_1 = (x_1, y_1)$$

$$P_2 = (x_2, y_2)$$

where the direction of the vector is from P_1 to P_2. Then the standard form of **a′**, denoted **a**, is defined by point P such that:

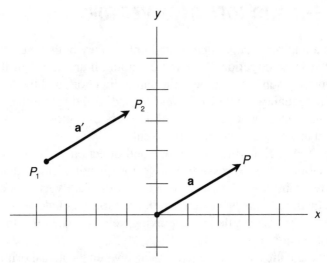

Fig. 12-5.   Standard and nonstandard forms of a vector in the *xy*-plane.

$$P = (x,y)$$
$$= [(x_2 - x_1),(y_2 - y_1)]$$

The two vectors **a** and **a'** are equivalent, because they have the same magnitude and the same direction, even though their originating points differ (Fig. 12-5).

IN THE POLAR PLANE

Vectors in polar coordinates are always denoted in standard form, that is, with their origins at $(\theta,r) = (0,0)$. This is necessary, because otherwise the vector direction becomes difficult to define.

IN *xyz*-SPACE

In three-dimensional Cartesian coordinates, suppose the originating and terminating points of a vector **a'** are P_1 and P_2, defined as follows:

$$P_1 = (x_1,y_1,z_1)$$
$$P_2 = (x_2,y_2,z_2)$$

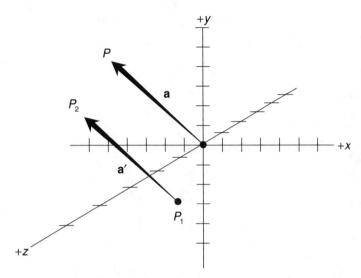

Fig. 12-6. Standard and nonstandard forms of a vector in xyz-space.

where the direction of the vector is from P_1 to P_2. Then the standard form of $\mathbf{a}'$, denoted $\mathbf{a}$, is defined by point P such that:

$$P = (x,y,z)$$
$$= [(x_2 - x_1),(y_2 - y_1),(z_2 - z_1)]$$

The two vectors $\mathbf{a}$ and $\mathbf{a}'$ are equivalent, because they have the same magnitude and the same direction, even though their originating points differ (Fig. 12-6).

Basic Properties

Here are some basic properties that apply to vectors and real-number scalars in the xy-plane, in the polar plane, or in xyz-space.

MULTIPLICATION OF VECTOR BY SCALAR

The following rules apply to scalars (that is, real-number quantities), and to vectors in standard form.

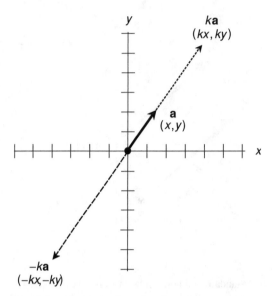

Fig. 12-7. Multiplication of a vector by a positive real scalar k, and by a negative real scalar $-k$, in the xy-plane.

In the xy-plane. In two-dimensional Cartesian coordinates, let vector **a** be defined by the coordinates (x,y) as shown in Fig. 12-7. Suppose **a** is multiplied by a positive real scalar k. Then the following equation holds:

$$k\mathbf{a} = k(x,y) = (kx,ky)$$

If **a** is multiplied by the scalar $k = 0$, then $k\mathbf{a} = \mathbf{0}$. If **a** is multiplied by a negative real scalar $-k$, then:

$$-k\mathbf{a} = -k(x,y) = (-kx,-ky)$$

In the polar plane. In two-dimensional polar coordinates, let vector **a** be defined by the coordinates (θ,r) as shown in Fig. 12-8. Suppose **a** is multiplied by a positive real scalar k. Then the following equation holds:

$$k\mathbf{a} = (\theta,kr)$$

If **a** is multiplied by the scalar $k = 0$, then $k\mathbf{a} = \mathbf{0}$. If **a** is multiplied by a negative real scalar $-k$, then:

$$-k\mathbf{a} = (\theta + \pi,kr)$$

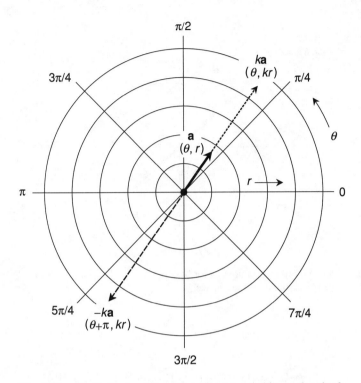

Fig. 12-8.   Multiplication of a vector (θ, r) by a positive real scalar k, and by a negative real scalar $-k$, in the polar plane. Note that the radius of a vector in polar coordinates is not allowed to be negative.

The addition of π rad (180°) to the angular coordinate is necessary because, in the polar-coordinate expression of a vector, the magnitude (and therefore the radius coordinate) is always positive or zero by convention.

In xyz-space. In three-dimensional Cartesian coordinates, let vector **a** be defined by the coordinates (x, y, z) as shown in Fig. 12-9. Suppose **a** is multiplied by a positive real scalar k. Then the following equation holds:

$$k\mathbf{a} = k(x, y, z) = (kx, ky, kz)$$

If **a** is multiplied by a negative real scalar $-k$, then:

$$-k\mathbf{a} = -k(x, y, z) = (-kx, -ky, -kz)$$

Suppose the direction angles of **a** are represented by $(\theta_x, \theta_y, \theta_z)$. Then the direction angles of $k\mathbf{a}$ are also given by $(\theta_x, \theta_y, \theta_z)$. The direction angles of $-k\mathbf{a}$ are the supplements of $(\theta_x, \theta_y, \theta_z)$, respectively. For direction angles in degrees:

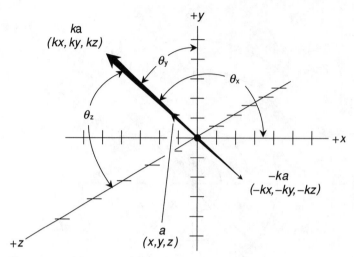

Fig. 12-9.   Multiplication of a vector by a positive real scalar k, and by a negative real scalar $-k$, in xyz-space.

$$\text{dir}\,(-k\mathbf{a}) = [(180° - \theta_x),(180° - \theta_y),(180° - \theta_z)]$$

For direction angles in radians:

$$\text{dir}\,(-k\mathbf{a}) = [(\pi - \theta_x),(\pi - \theta_y),(\pi - \theta_z)]$$

Suppose the direction cosines of $\mathbf{a}$ are represented by (α,β,γ). Then the direction cosines of $k\mathbf{a}$ are also given by (α,β,γ). The direction cosines of $-k\mathbf{a}$ are the negatives of α, β, and γ, respectively:

$$\text{dir}\,(-k\mathbf{a}) = (-\alpha,-\beta,-\gamma)$$

UNIT VECTORS

Consider two vectors $\mathbf{a}$ and $\mathbf{b}$ expressed in standard form in Cartesian 3-space. Suppose $\mathbf{a} = (x_a,y_a,z_a)$ and $\mathbf{b} = (x_b,y_b,z_b)$. Either of these two vectors can be broken down into a sum of three mutually perpendicular *component vectors*, each of which lies along one of the coordinate axes. These component vectors are, in turn, each scalar multiples of mutually perpendicular vectors with magnitude 1, as follows:

$$\mathbf{a} = (x_a, y_a, z_a)$$
$$= (x_a, 0, 0) + (0, y_a, 0) + (0, 0, z_a)$$
$$= x_a(1, 0, 0) + y_a(0, 1, 0) + z_a(0, 0, 1)$$

$$\mathbf{b} = (x_b, y_b, z_b)$$
$$= (x_b, 0, 0) + (0, y_b, 0) + (0, 0, z_b)$$
$$= x_b(1, 0, 0) + y_b(0, 1, 0) + z_b(0, 0, 1)$$

The three mutually perpendicular vectors $(1,0,0)$, $(0,1,0)$, and $(0,0,1)$ are called *unit vectors* because they all have magnitude 1. It is customary to name these three standard unit vectors **i**, **j**, and **k**, like this:

$$(1,0,0) = \mathbf{i}$$

$$(0,1,0) = \mathbf{j}$$

$$(0,0,1) = \mathbf{k}$$

The vectors **a** and **b**, and their sum vector **a** + **b** (as shown in Fig. 12-3, for example), break down this way:

$$\mathbf{a} = (x_a, y_a, z_a)$$
$$= x_a\mathbf{i} + y_a\mathbf{j} + z_a\mathbf{k}$$

$$\mathbf{b} = (x_b, y_b, z_b)$$
$$= x_b\mathbf{i} + y_b\mathbf{j} + z_b\mathbf{k}$$

$$\mathbf{a} + \mathbf{b} = (x_a + x_b)\mathbf{i} + (y_a + y_b)\mathbf{j} + (z_a + z_b)\mathbf{k}$$

PROBLEM 12-10
Break the vector $\mathbf{b} = (-2,3,-7)$ down into a sum of multiples of the unit vectors **i**, **j**, and **k**.

SOLUTION 12-10
If you have trouble envisioning this situation, imagine **i** as having 1 unit of "width going to the right," **j** as having 1 unit of "altitude going straight up," and **k** as having 1 unit of "depth coming straight at you." The breakdown proceeds like this:

$$\mathbf{b} = (-2,3,-7)$$
$$= -2 \times (1,0,0) + 3 \times (0,1,0) + [-7 \times (0,0,1)]$$
$$= -2\mathbf{i} + 3\mathbf{j} + (-7)\mathbf{k}$$
$$= -2\mathbf{i} + 3\mathbf{j} - 7\mathbf{k}$$

COMMUTATIVITY OF VECTOR ADDITION

When summing any two vectors, it does not matter in which order the addition is done. The resultant vector is the same in either case. If **a** and **b** are vectors, then:

$$\mathbf{a} + \mathbf{b} = \mathbf{b} + \mathbf{a}$$

COMMUTATIVITY OF VECTOR-SCALAR MULTIPLICATION

When a vector is multiplied by a scalar, it does not matter in which order the multiplication is done. The resultant vector is the same in either case. If **a** is a vector and k is a scalar, then:

$$k\mathbf{a} = \mathbf{a}k$$

COMMUTATIVITY OF DOT PRODUCT

When the dot product of two vectors is determined, it does not matter in which order the operation is performed. The result is the same scalar quantity in either case. If **a** and **b** are vectors, then:

$$\mathbf{a} \bullet \mathbf{b} = \mathbf{b} \bullet \mathbf{a}$$

NEGATIVE COMMUTATIVITY OF CROSS PRODUCT

Let θ_{ab} be the angle between two vectors **a** and **b** as defined in the plane containing **a** and **b**, such that $0° \le \theta_{ab} \le 180°$ ($0 \le \theta_{ab} \le \pi$), and such that the rotational sense of the angle is not considered. In this case, the magnitude of the cross product vector is a nonnegative real number, and is independent of the order in

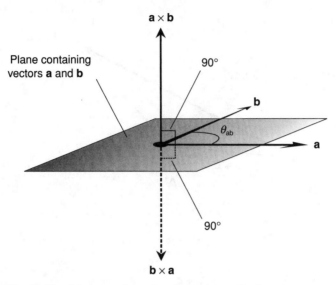

Fig. 12-10.   The vector **b** × **a** has the same magnitude as vector
a × **b**, but points in the opposite direction.

which the operation is performed. This can be proven on the basis of the commutative property for multiplication of scalars, a fact of ordinary arithmetic, as follows:

$$|\mathbf{a} \times \mathbf{b}| = |\mathbf{a}|\ |\mathbf{b}|\ \sin \theta_{ab}$$

$$|\mathbf{b} \times \mathbf{a}| = |\mathbf{b}|\ |\mathbf{a}|\ \sin \theta_{ab} = |\mathbf{a}|\ |\mathbf{b}|\ \sin \theta_{ab}$$

The direction of **b** × **a** is opposite that of **a** × **b**. This is shown in the example of Fig. 12-10. Therefore:

$$\mathbf{b} \times \mathbf{a} = (-1)(\mathbf{a} \times \mathbf{b}) = -(\mathbf{a} \times \mathbf{b})$$

ASSOCIATIVITY OF VECTOR ADDITION

When summing vectors, it makes no difference how the addends are grouped. The resultant vector is always the same. The case for three vectors is easily stated. If **a**, **b**, and **c** are vectors, then:

$$(\mathbf{a} + \mathbf{b}) + \mathbf{c} = \mathbf{a} + (\mathbf{b} + \mathbf{c})$$

This situation is shown in Fig. 12-11. Drawing A shows the process of summing the three vectors as (**a** + **b**) + **c**. Drawing B shows the process of summing them as **a** + (**b** + **c**).

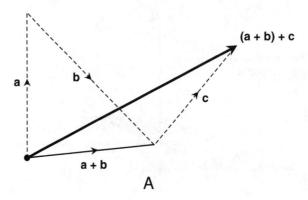

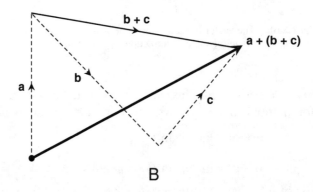

Fig. 12-11. At A, the sum of three vectors expressed as $(\mathbf{a} + \mathbf{b}) + \mathbf{c}$. At B, the sum of the same three vectors expressed as $\mathbf{a} + (\mathbf{b} + \mathbf{c})$.

ASSOCIATIVITY OF VECTOR-SCALAR MULTIPLICATION

Let k_1 and k_2 be scalar quantities, and let $\mathbf{a}$ be a vector quantity. If all three quantities are multiplied, it makes no difference how they are grouped; the resultant vector is always the same. The following equation applies:

$$k_1(k_2\mathbf{a}) = (k_1 k_2)\mathbf{a}$$

Other Properties

Here are some more properties that apply to vectors and real-number scalars. You'll encounter these facts less often than the ones in the preceding section, but you might find them useful when manipulating vector and scalar quantities in physics and engineering.

DISTRIBUTIVITY OF SCALAR MULTIPLICATION OVER SCALAR ADDITION

Let $\mathbf{a}$ be a vector, and let k_1 and k_2 be real-number scalars. Then the following equations hold. The resultant is a vector quantity:

$$(k_1 + k_2)\mathbf{a} = k_1\mathbf{a} + k_2\mathbf{a}$$

$$\mathbf{a}(k_1 + k_2) = \mathbf{a}k_1 + \mathbf{a}k_2$$
$$= k_1\mathbf{a} + k_2\mathbf{a}$$

DISTRIBUTIVITY OF SCALAR MULTIPLICATION OVER VECTOR ADDITION

Let $\mathbf{a}$ and $\mathbf{b}$ be vectors, and let k be a real-number scalar. Then the following equations hold. The resultant is a vector quantity:

$$k(\mathbf{a} + \mathbf{b}) = k\mathbf{a} + k\mathbf{b}$$

$$(\mathbf{a} + \mathbf{b})k = \mathbf{a}k + \mathbf{b}k = k\mathbf{a} + k\mathbf{b}$$

DISTRIBUTIVITY OF DOT PRODUCT OVER VECTOR ADDITION

Let $\mathbf{a}$, $\mathbf{b}$, and $\mathbf{c}$ be vectors. Then the following equations hold. The resultant is a scalar quantity:

$$\mathbf{a} \bullet (\mathbf{b} + \mathbf{c}) = (\mathbf{a} \bullet \mathbf{b}) + (\mathbf{a} \bullet \mathbf{c})$$

$$(\mathbf{b} + \mathbf{c}) \bullet \mathbf{a} = (\mathbf{b} \bullet \mathbf{a}) + (\mathbf{c} \bullet \mathbf{a})$$
$$= (\mathbf{a} \bullet \mathbf{b}) + (\mathbf{a} \bullet \mathbf{c})$$

DISTRIBUTIVITY OF CROSS PRODUCT OVER VECTOR ADDITION

Let $\mathbf{a}$, $\mathbf{b}$, and $\mathbf{c}$ be vectors. Then the following equation holds. The resultant is a vector quantity:

$$\mathbf{a} \times (\mathbf{b} + \mathbf{c}) = (\mathbf{a} \times \mathbf{b}) + (\mathbf{a} \times \mathbf{c})$$

This property can be expanded to apply to pairs of vector sums, each having n addends (where $n = 2$, $n = 3$, $n = 4$, etc.), in the same way as multiplication is distributive with respect to addition in real-number arithmetic. For example, for $n = 2$:

$$(\mathbf{a} + \mathbf{b}) \times (\mathbf{c} + \mathbf{d}) = (\mathbf{a} \times \mathbf{c}) + (\mathbf{a} \times \mathbf{d}) + (\mathbf{b} \times \mathbf{c}) + (\mathbf{b} \times \mathbf{d})$$

In the case of $n = 3$, the cross product of two vector sums expands like this:

$$(\mathbf{a} + \mathbf{b} + \mathbf{c}) \times (\mathbf{d} + \mathbf{e} + \mathbf{f}) = (\mathbf{a} \times \mathbf{d}) + (\mathbf{a} \times \mathbf{e}) + (\mathbf{a} \times \mathbf{f})$$
$$+ (\mathbf{b} \times \mathbf{d}) + (\mathbf{b} \times \mathbf{e}) + (\mathbf{b} \times \mathbf{f})$$
$$+ (\mathbf{c} \times \mathbf{d}) + (\mathbf{c} \times \mathbf{e}) + (\mathbf{c} \times \mathbf{f})$$

DOT PRODUCT OF CROSS PRODUCTS

Let $\mathbf{a}$, $\mathbf{b}$, $\mathbf{c}$, and $\mathbf{d}$ be vectors. Then the following equation holds. The resultant is a scalar quantity:

$$(\mathbf{a} \times \mathbf{b}) \bullet (\mathbf{c} \times \mathbf{d}) = (\mathbf{a} \bullet \mathbf{c})(\mathbf{b} \bullet \mathbf{d}) - (\mathbf{a} \bullet \mathbf{d})(\mathbf{b} \bullet \mathbf{c})$$

DOT PRODUCT OF MIXED VECTORS AND SCALARS

Let t and u be scalars, and let $\mathbf{a}$ and $\mathbf{b}$ be vectors. Then the following equation holds. The resultant is a scalar quantity:

$$t\mathbf{a} \bullet u\mathbf{b} = tu(\mathbf{a} \bullet \mathbf{b})$$

CROSS PRODUCT OF MIXED VECTORS AND SCALARS

Let t and u be scalars, and let $\mathbf{a}$ and $\mathbf{b}$ be vectors. Then the following equation holds. The resultant is a vector quantity:

$$t\mathbf{a} \times u\mathbf{b} = tu(\mathbf{a} \times \mathbf{b})$$

PROBLEM 12-11

Consider two vectors, $\mathbf{a} = (x_a, y_a, z_a)$ and $\mathbf{b} = (x_b, y_b, z_b)$. Derive a general formula for their cross product in terms of x, y, and z coordinates. That is, produce an expression for $\mathbf{a} \times \mathbf{b}$ in the form of an ordered triple.

SOLUTION 12-11

Recall the concept of the unit vectors $\mathbf{i} = (1,0,0)$, $\mathbf{j} = (0,1,0)$, and $\mathbf{k} = (0,0,1)$. Note the following facts, which can be verified using the right-hand rule and the formula for the magnitude of the cross product of vectors:

$$\mathbf{i} \times \mathbf{j} = \mathbf{k}$$

$$\mathbf{j} \times \mathbf{i} = -\mathbf{k}$$

$$\mathbf{i} \times \mathbf{k} = -\mathbf{j}$$

$$\mathbf{k} \times \mathbf{i} = \mathbf{j}$$

$$\mathbf{j} \times \mathbf{k} = \mathbf{i}$$

$$\mathbf{k} \times \mathbf{j} = -\mathbf{i}$$

Note the distributive property of the cross product over vector addition, extrapolated for the cross product of two vector sums. Also, note the way scalar and vector products mix. Converting $\mathbf{a}$ and $\mathbf{b}$ to sums of unit vectors, we have:

$$\mathbf{a} \times \mathbf{b} = (x_a, y_a, z_a) \times (x_b, y_b, z_b)$$
$$= (x_a\mathbf{i} + y_a\mathbf{j} + z_a\mathbf{k}) \times (x_b\mathbf{i} + y_b\mathbf{j} + z_b\mathbf{k})$$

Using the distributive property of the cross product with respect to vector addition, we can expand this, as follows:

$$\mathbf{a} \times \mathbf{b} = (x_a\mathbf{i} \times x_b\mathbf{i}) + (x_a\mathbf{i} \times y_b\mathbf{j}) + (x_a\mathbf{i} \times z_b\mathbf{k})$$
$$+ (y_a\mathbf{j} \times x_b\mathbf{i}) + (y_a\mathbf{j} \times y_b\mathbf{j}) + (y_a\mathbf{j} \times z_b\mathbf{k})$$
$$+ (z_a\mathbf{k} \times x_b\mathbf{i}) + (z_a\mathbf{k} \times y_b\mathbf{j}) + (z_a\mathbf{k} \times z_b\mathbf{k})$$

Using our knowledge of how scalar multiplication and cross products can be mixed, the above can be rewritten to obtain:

$$\mathbf{a} \times \mathbf{b} = x_a x_b(\mathbf{i} \times \mathbf{i}) + x_a y_b(\mathbf{i} \times \mathbf{j}) + x_a z_b(\mathbf{i} \times \mathbf{k})$$
$$+ y_a x_b(\mathbf{j} \times \mathbf{i}) + y_a y_b(\mathbf{j} \times \mathbf{j}) + y_a z_b(\mathbf{j} \times \mathbf{k})$$
$$+ z_a x_b(\mathbf{k} \times \mathbf{i}) + z_a y_b(\mathbf{k} \times \mathbf{j}) + z_a z_b(\mathbf{k} \times \mathbf{k})$$

Remember that the cross product of any vector with itself is the zero vector, and that the zero vector times any scalar is the zero vector. We can therefore rewrite the above in this form:

$$\mathbf{a} \times \mathbf{b} = 0 + x_a y_b (\mathbf{i} \times \mathbf{j}) + x_a z_b (\mathbf{i} \times \mathbf{k})$$
$$+ y_a x_b (\mathbf{j} \times \mathbf{i}) + 0 + y_a z_b (\mathbf{j} \times \mathbf{k})$$
$$+ z_a x_b (\mathbf{k} \times \mathbf{i}) + z_a y_b (\mathbf{k} \times \mathbf{j}) + 0$$

Taking note of the six facts given above for the cross products of unit vectors $\mathbf{i}$, $\mathbf{j}$, and $\mathbf{k}$, and getting rid of the 0 vectors in the equation, the above can be simplified to obtain:

$$\mathbf{a} \times \mathbf{b} = x_a y_b \mathbf{k} + x_a z_b (-\mathbf{j})$$
$$+ y_a x_b (-\mathbf{k}) + y_a z_b \mathbf{i}$$
$$+ z_a x_b \mathbf{j} + z_a y_b (-\mathbf{i})$$

Rearranging the signs, we get this:

$$\mathbf{a} \times \mathbf{b} = x_a y_b \mathbf{k} - x_a z_b \mathbf{j}$$
$$- y_a x_b \mathbf{k} + y_a z_b \mathbf{i}$$
$$+ z_a x_b \mathbf{j} - z_a y_b \mathbf{i}$$

This can be rewritten yet again, based on the rules for vector sums:

$$\mathbf{a} \times \mathbf{b} = (y_a z_b - z_a y_b)\mathbf{i} + (z_a x_b - x_a z_b)\mathbf{j} + (x_a y_b - y_a x_b)\mathbf{k}$$

Therefore, the general formula for $\mathbf{a} \times \mathbf{b}$ as an ordered triple is as follows:

$$\mathbf{a} \times \mathbf{b} = [(y_a z_b - z_a y_b), (z_a x_b - x_a z_b), (x_a y_b - y_a x_b)]$$

Quick Practice

Here are some practice problems that cover the material presented in this chapter. Solutions follow the problems.

PROBLEMS

1. Find the magnitude of the vector $\mathbf{a} = (-7, -10)$ in the xy-plane. Assume the values given are exact. Express the answer to three significant figures.

2. Convert the vector $\mathbf{a} = (-7,-10)$, as expressed in the xy-plane, to polar form as a vector $\mathbf{a} = (\theta_a, r_a)$. Assume the values given are exact. Express the answer to three significant figures, with θ_a in degrees.

3. Find the magnitude of the vector $\mathbf{b} = (8, -1,-6)$ in xyz-space. Assume the values given are exact. Express the answer to four significant figures.

4. Consider the two vectors $\mathbf{a} = (-7,-10,0)$ and $\mathbf{b} = (8,-1,-6)$ in xyz-space. What is their dot product?

5. Consider the two vectors $\mathbf{a} = (2,6,0)$ and $\mathbf{b} = (7,4,3)$ in xyz-space. What is their cross product?

SOLUTIONS

1. Use the formula for vector magnitude in the Cartesian plane:

$$|\mathbf{a}| = (x_a^2 + y_a^2)^{1/2}$$
$$= [(-7)^2 + (-10)^2]^{1/2}$$
$$= (49 + 100)^{1/2}$$
$$= 149^{1/2}$$
$$= 12.2$$

2. First, find the angle θ_a according to the formula for the direction of a vector $\mathbf{a}$ in the Cartesian plane when $x_a < 0$:

$$\text{dir } \mathbf{a} = \theta_a$$
$$= 180° + \arctan (y_a/x_a)$$
$$= 180° + \arctan (-10/-7)$$
$$= 180° + 55.0°$$
$$= 235°$$

We found r_a in the previous problem. It is $|\mathbf{a}|$, which is 12.2. In polar coordinates:

$$\mathbf{a} = (\theta_a, r_a)$$
$$= (235°, 12.2)$$

3. Use the formula for vector magnitude in Cartesian 3-space:

$$|\mathbf{a}| = (x_a^2 + y_a^2 + z_a^2)^{1/2}$$
$$= [8^2 + (-1)^2 + (-6)^2]^{1/2}$$
$$= (64 + 1 + 36)^{1/2}$$
$$= 101^{1/2}$$
$$= 10.05$$

4. Use the formula for the dot product of two vectors in Cartesian 3-space. Extra brackets are added for clarity:

$$\mathbf{a} \bullet \mathbf{b} = x_a x_b + y_a y_b + z_a z_b$$
$$= (-7 \times 8) + [(-10) \times (-1)[+ [0 \times (-6)]$$
$$= -56 + 10 + 0$$
$$= -46$$

5. Use the formula for the cross product of two vectors in Cartesian 3-space, derived in Solution 12-11 above. Extra brackets and braces are added for clarity:

$$\mathbf{a} \times \mathbf{b} = [(y_a z_b - z_a y_b),(z_a x_b - x_a z_b),(x_a y_b - y_a x_b)]$$
$$= \{[(6 \times 3) - (0 \times 4)],[(0 \times 7) - (2 \times 3)],[(2 \times 4) - (6 \times 7)]\}$$
$$= [(18 - 0),(0 - 6),(8 - 42)]$$
$$= (18,-6,-34)$$

Quiz

This is an "open book" quiz. You may refer to the text in this chapter. A good score is 8 correct. Answers are in the back of the book.

1. Imagine the vector $\mathbf{a} = (x_a, y_a, z_a) = (1,1,1)$ in Cartesian 3-space. Which of the following statements about $\mathbf{a}$ is true?

   (a) Vector $\mathbf{a}$ points right along the x axis.
   (b) Vector $\mathbf{a}$ points right along the y axis.
   (c) Vector $\mathbf{a}$ points right along the z axis.
   (d) Vector $\mathbf{a}$ does not point along the x, y, or z axis.

2. Imagine the vector $\mathbf{a} = (x_a, y_a, z_a) = (1,1,1)$ in Cartesian 3-space. What is dir $\mathbf{a}$, specified in the form of an ordered triple of direction angles $(\theta_x, \theta_y, \theta_z)$ relative to the x, y, and z axes, respectively? Express the angles in degrees, to three significant figures. (Warning: This is a tricky problem. These angles are not 45°, as you might at first suppose. Consider vector $\mathbf{a}$ as the *internal diagonal of a cube* that is 1 unit high, 1 unit wide, and 1 unit deep.)

   (a) (54.7°,54.7°,54.7°)
   (b) (35.3°,35.3°,35.3°)

(c) $(60.0°,60.0°,60.0°)$

(d) $(30.0°,30.0°,30.0°)$

3. Imagine the vector $\mathbf{a} = (x_a,y_a,z_a) = (1,1,1)$ in Cartesian 3-space. What is dir $\mathbf{a}$, specified in the form of an ordered triple of direction cosines (α,β,γ)? Express the cosines to two significant figures.

(a) $(0.50,0.50,0.50)$

(b) $(0.58,0.58.0.58)$

(c) $(0.82,0.82,0.82)$

(d) $(0.87,0.87,0.87)$

4. Consider a vector $\mathbf{a} = (x_a,y_a)$ in the Cartesian plane. Suppose each of these values is multiplied by 2, producing the resultant vector $\mathbf{b} = (2x_a,2y_a)$. The value of $|\mathbf{b}|$ is

(a) equal to $2|\mathbf{a}|$.

(b) equal to $4|\mathbf{a}|$.

(c) equal to $8|\mathbf{a}|$.

(d) impossible to determine without more information.

5. Consider a vector $\mathbf{a} = (x_a,y_a,z_a)$ in Cartesian 3-space. Suppose each of these values is multiplied by 2, producing the resultant vector $\mathbf{b} = (2x_a,2y_a,2z_a)$. The value of $|\mathbf{b}|$ is

(a) equal to $2|\mathbf{a}|$.

(b) equal to $4|\mathbf{a}|$.

(c) equal to $8|\mathbf{a}|$.

(d) impossible to determine without more information.

6. Suppose you are told that a certain vector is defined as $\mathbf{d} = 3\mathbf{i} - 4\mathbf{j} + 0\mathbf{k}$ in xyz-space. From this, you can be certain that the vector

(a) is perpendicular to the $+x$ axis.

(b) is perpendicular to the $+y$ axis.

(c) is perpendicular to the $+z$ axis.

(d) is not perpendicular to any of the three axes $+x$, $+y$, or $+z$.

7. Suppose you are told that the direction angles in xyz-space for vector $\mathbf{h}$ are $(30°,60°,90°)$. What are the direction angles for the vector $-2\mathbf{h}$?

(a) $(-30°,-60°,-90°)$.

(b) $(150°,120°,90°)$.

(c) $(60°,120°,180°)$.

(d) None of the above

8. In *xyz*-space, the sum of any vector with another vector having equal magnitude but opposite direction is equal to
   (a) the scalar quantity 0.
   (b) the vector **0**.
   (c) a vector with twice the magnitude of, but the same direction as, the original.
   (d) a quantity that requires more information to be defined.

9. In physics, it is a well-known fact that the force **F** (a vector) exerted on an object is equal to the product of the mass *m* (a positive real-number scalar) of the object multiplied by the acceleration **a** (a vector) of the object. Mathematically:

$$\mathbf{F} = m\mathbf{a}$$

Based on this, we can be certain that

   (a) **F** and **a** are perpendicular to each other.
   (b) **F** and **a** point in opposite directions.
   (c) **F** and **a** point in the same direction.
   (d) **F** and **a** have supplementary direction cosines.

10. The cross product of two vectors is
   (a) always a scalar.
   (b) never a scalar.
   (c) sometimes a scalar.
   (d) always negative or zero.

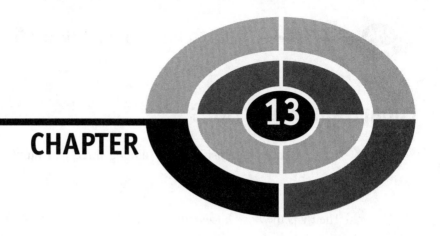

Logarithmic and Exponential Functions

A *logarithm* (sometimes called a *log*) of a quantity is an exponent to which a positive real-number constant must be raised to obtain that quantity. An *exponential* of a quantity is the result of raising a positive real-number constant to a power equal to that quantity. The constant is known as the *base* of the *logarithmic function* or the *exponential function*. The two most common bases for logarithmic and exponential functions are 10 and *e*, where *e* is an irrational number approximately equal to 2.71828. The number *e* is also known as *Euler's constant* and the *exponential constant*.

Logarithmic Functions

Suppose this relationship exists among three real numbers b, x, and y, where $b > 0$:

$$b^y = x$$

Then y is the *base b logarithm* of x. This expression is written

$$y = \log_b x$$

COMMON LOGARITHMS

Base-10 logarithms are also known as *common logarithms* or *common logs*. In equations, common logarithms are denoted by writing "log" without a subscript. For example:

$$\log 10 = 1.000$$

Figure 13-1 is a partial linear-coordinate graph of the function $y = \log x$. Figure 13-2 is a partial graph of the same function in semilog coordinates.

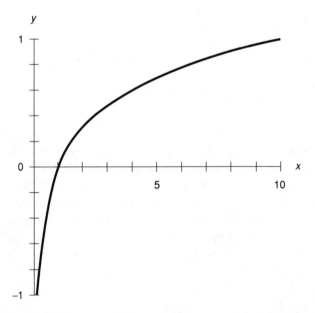

Fig. 13-1. Partial linear-coordinate graph of the common logarithm function. As x approaches 0, the value of y becomes arbitrarily large in the negative sense.

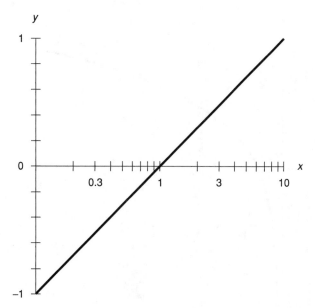

Fig. 13-2. Partial semilog-coordinate graph of the common logarithm function. In this illustration, the y axis intersects the x axis at the point where $x = 0.1$.

A logarithmic function of x, regardless of the base, is defined only when $x > 0$. The set of x values for which any function of x is defined is called the *domain* of the function. The resulting y (or output) values encompass the entire set of real numbers. The set of all possible y values in a function is known as the *range* of the function.

NATURAL LOGARITHMS

Base e logarithms are also called *natural logs* or *Napierian logs*. In equations, the natural-log function is usually denoted "ln" or "$\log_e$." For example:

$$\ln 2.71828 = \log_e 2.71828 \approx 1.00000$$

Figure 13-3 is a partial linear-coordinate graph of the function $y = \ln x$. Figure 13-4 is a partial graph of the same function in semilog coordinates. As with the base-10 logarithmic function, the domain is limited to the set of positive real numbers, but the range is the set of all real numbers.

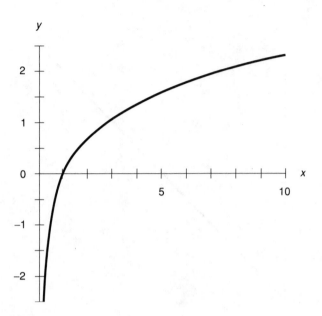

Fig. 13-3. Partial linear-coordinate graph of the natural logarithm function. As *x* approaches 0, the value of *y* becomes arbitrarily large in the negative sense.

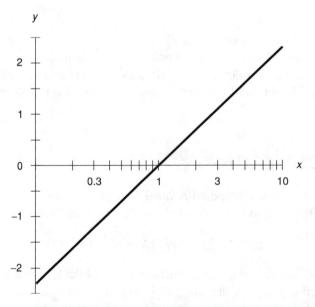

Fig. 13-4. Partial semilog-coordinate graph of the natural logarithm function. In this illustration, the *y* axis intersects the *x* axis at the point where *x* = 0.1.

How Logarithmic Functions Behave

Here are some important properties of logarithms that apply to any positive real-number base, including 10 and e. The principles applicable to logarithms of products, ratios, and powers are worth memorizing. In the following rules, let "$\log_b$" refer to a logarithm of any positive real-number base b (even including $b = 1$, although that case is trivial).

LOGARITHM OF PRODUCT

Let x and y be positive real numbers. The logarithm of the product is equal to the sum of the logarithms of the individual numbers:

$$\log_b xy = \log_b x + \log_b y$$

LOGARITHM OF RATIO

Let x and y be positive real numbers. The logarithm of their ratio, or quotient, is equal to the difference between the logarithms of the individual numbers:

$$\log_b (x/y) = \log_b x - \log_b y$$

LOGARITHM OF POWER

Let x be a positive real number. Let y be any real number (positive, negative, or zero). The logarithm of x raised to the power y can be reduced to a product, as follows:

$$\log_b x^y = y \log_b x$$

LOGARITHM OF RECIPROCAL

Let x be a positive real number. The logarithm of the reciprocal (or *multiplicative inverse*) of x is equal to the negative (or *additive inverse*) of the logarithm of x, as follows:

$$\log_b (1/x) = -\log_b x$$

LOGARITHM OF ROOT

Let x be a positive real number. Let y be any real number except zero. The logarithm of the yth root of x (also denoted as x to the $1/y$ power) is given by:

$$\log_b (x^{1/y}) = (\log_b x)/y$$

LOGARITHM OF THE BASE RAISED TO A POWER

The base b logarithm of b raised to any real-number power is always equal to that real number. Thus, for every x, the following equation holds:

$$\log_b (b^x) = x$$

COMMON LOGARITHM IN TERMS OF NATURAL LOGARITHM

Let x be a positive real number. The common logarithm of x can be expressed in terms of the natural logarithms of x and 10, as follows:

$$\log_{10} x = (\ln x)/(\ln 10) \approx 0.434 \ln x$$

NATURAL LOGARITHM IN TERMS OF COMMON LOGARITHM

Let x be a positive real number. The natural logarithm of x can be expressed in terms of the common logarithms of x and e, as follows:

$$\ln x = (\log_{10} x)/(\log_{10} e) \approx 2.303 \log_{10} x$$

PROBLEM 13-1
Compare the common logarithms of 0.01, 0.1, 1, 10, and 100.

SOLUTION 13-1
The common logarithm of a number is the power of 10 that produces that number. Note that $0.01 = 10^{-2}$, $0.1 = 10^{-1}$, $1 = 10^0$, $10 = 10^1$, and $100 = 10^2$. Therefore:

$$\log 0.01 = -2$$

$$\log 0.1 = -1$$

$$\log 1 = 0$$

$$\log 10 = 1$$

$$\log 100 = 2$$

PROBLEM 13-2

What is the number whose common logarithm is 15? What is the number whose common logarithm is −15?

SOLUTION 13-2

The number whose common logarithm is 15 is the quantity 10^{15} (or 1,000,000,000,000,000). In spoken English, this number is called a *quadrillion*. The number whose common logarithm is −15 is the quantity 10^{-15} (or 0.000000000000001). In spoken English, this would be called a *quadrillionth*.

PROBLEM 13-3

Compare the base-*e* logarithms (or natural logarithms) of 0.01, 0.1, 1, 10, and 100. Assume the input values are exact, and express the answers to four significant figures.

SOLUTION 13-3

The base-*e* logarithm of a number is the power of *e* that produces that number. These are best found using a calculator. The results are as follows, to four significant figures in each case:

$$\ln 0.01 = -4.605$$

$$\ln 0.1 = -2.303$$

$$\ln 1 = 0.000$$

$$\ln 10 = 2.303$$

$$\ln 100 = 4.605$$

PROBLEM 13-4

Why is the logarithm of 0, or of any negative number, not defined in the set of real numbers?

SOLUTION 13-4

Let's see what happens if we try to calculate the base b logarithm of -2. Remember that b is a positive real number. Suppose $\log_b -2 = y$. This can be rewritten in the form $b^y = -2$. No real number y satisfies this equation. No matter what y might happen to be, the value of b^y is always positive. If we change -2 to any other negative number, or to 0, in this scenario, the same problem occurs. It's impossible to find any real number y, such that b^y is less than or equal to 0.

Exponential Functions

Suppose this relationship exists among three real numbers b, x, and y, where $b > 0$:

$$b^x = y$$

Then y is the *base b exponential* of x. The two most common exponential function bases are $b = 10$ and $b = e \approx 2.71828$.

COMMON EXPONENTIALS

Base 10 exponentials are also known as *common exponentials*. For example:

$$10^{-3.000} = 0.001$$

Figure 13-5 is a partial linear-coordinate graph of the function $y = 10^x$. Figure 13-6 is a partial graph of the same function in semilog coordinates. The domain encompasses the entire set of real numbers. The range is limited to the positive real numbers.

NATURAL EXPONENTIALS

Base e exponentials are also known as *natural exponentials*. For example:

$$e^{-3.000} \approx 2.71828^{-3.000} \approx 0.04979$$

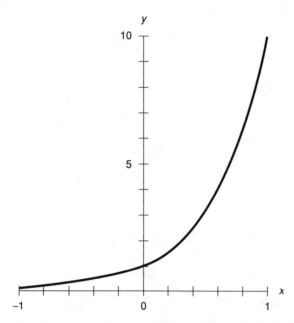

Fig. 13-5. Partial linear-coordinate graph of the common exponential function. As x becomes arbitrarily large in the negative sense, the value of y approaches 0.

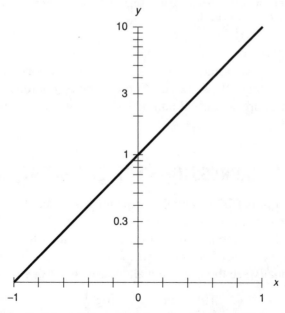

Fig. 13-6. Partial semilog graph of the common exponential function. In this illustration, the x axis intersects the y axis at the point where $y = 0.1$.

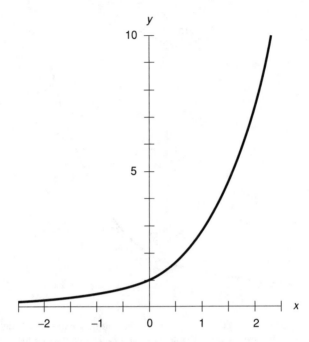

Fig. 13-7. Partial linear-coordinate graph of the natural exponential function. As x becomes arbitrarily large in the negative sense, the value of y approaches 0.

Figure 13-7 is a partial linear-coordinate graph of the function $y = e^x$. Figure 13-8 is a partial graph of the same function in semilog coordinates. The domain encompasses the entire set of real numbers. The range is limited to the positive real numbers.

ALTERNATIVE EXPRESSIONS FOR EXPONENTIALS

Sometimes, the common exponential of a quantity is called the *common antilogarithm* (antilog) or the *common inverse logarithm* ($\log^{-1}$) of that number. Similarly, the natural exponential of a quantity is called the *natural antilogarithm* (antiln) or the *natural inverse logarithm* ($\ln^{-1}$) of that number. Therefore, you will occasionally see the following alternative notations:

$$10^x = \text{antilog } x = \log^{-1} x$$

$$e^x = \text{antiln } x = \ln^{-1} x$$

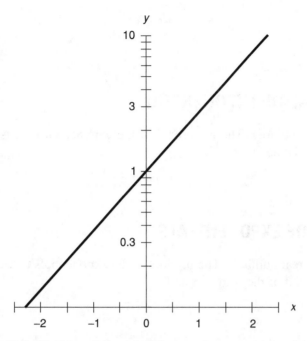

Fig. 13-8. Partial semilog graph of the natural exponential function. In this illustration, the x axis intersects the y axis at the point where $y = 0.1$.

How Exponential Functions Behave

Here are some important properties of exponential functions. As with the facts involving logarithms, you'll find it handy to memorize the principles applicable to products, ratios, and powers. In all the following formulas, let b be a positive real-number exponential base.

LOGARITHMS VERSUS EXPONENTIALS

The exponential function is the inverse of the logarithm function, and vice-versa. This means that two functions "undo" each other, provided both functions are defined for the values in question. For any real number x, and for any positive real number y, the following equations hold. In this particular example, let $\log_b$ be represented by lb (for simplicity of notation, avoiding the need for putting a subscript within a superscript!). Then:

$$\text{lb } (b^x) = x$$

$$b^{(\text{lb } y)} = y$$

RECIPROCAL OF EXPONENTIAL

Let x be a real number. The reciprocal of the exponential of x is equal to the exponential of the negative of x:

$$1/(b^x) = b^{-x}$$

PRODUCT OF EXPONENTIALS

Let x and y be real numbers. The product of the exponentials of x and y is equal to the exponential of the sum of x and y:

$$b^x b^y = b^{(x+y)}$$

EXPONENTIAL OF RATIO

Let x and y be real numbers, with the restriction $y \neq 0$. The exponential of x/y is equal to the exponential of $1/y$ with base b^x:

$$b^{(x/y)} = (b^x)^{(1/y)}$$

RATIO OF EXPONENTIALS

Let x and y be real numbers. The ratio of the exponential of x to the exponential of y is equal to the exponential of the difference between x and y:

$$b^x/b^y = b^{(x-y)}$$

EXPONENTIAL OF EXPONENTIAL

Let x and y be real numbers. The yth power of the exponential of x is equal to the exponential of the product xy:

$$(b^x)^y = b^{(xy)}$$

PRODUCT OF COMMON AND NATURAL EXPONENTIALS

Let x be a real number. The product of the common and natural exponentials of x is equal to the exponential of x to the base $10e$:

$$(10^x)(e^x) = (10e)^x \approx (27.1828)^x$$

RATIO OF COMMON TO NATURAL EXPONENTIAL

Let x be a real number. The ratio of the common exponential of x to the natural exponential of x is equal to the exponential of x to the base $10/e$:

$$10^x/e^x = (10/e)^x \approx (3.6788)^x$$

RATIO OF NATURAL TO COMMON EXPONENTIAL

Let x be a real number. The ratio of the natural exponential of x to the common exponential of x is equal to the exponential of x to the base $e/10$:

$$e^x/10^x = (e/10)^x \approx (0.271828)^x$$

PROBLEM 13-5
Compare the values of e^{-2}, e^{-1}, e^0, e^1, and e^2. Assume the exponents given here are exact. Express each answer to five significant figures.

SOLUTION 13-5
In order to determine the values of natural exponentials, it is necessary to use a calculator that has this function. The key is often labeled e^x. With some calculators, it is necessary to hit an "Inv" key followed by a "ln" key. Here are the values of the above exponentials, rounded off to five significant figures:

$$e^{-2} = 0.13534$$
$$e^{-1} = 0.36788$$
$$e^0 = 1.0000$$
$$e^1 = 2.7183$$
$$e^2 = 7.3891$$

PROBLEM 13-6

What is the number whose common exponential function value is 1,000,000? What is the number whose common exponential function value is 0.0001? Assume the values given are exact.

SOLUTION 13-6

The number 6 produces the common exponential value 1,000,000. This can be demonstrated by the fact that $10^6 = 1,000,000$. The number −4 produces the common exponential value 0.0001. This is shown by the fact that $10^{-4} = 0.0001$.

PROBLEM 13-7

What is the number whose natural exponential function value is 1,000,000? What is the number whose natural exponential function value is 0.0001? Assume the input values given here are exact. Express the answers to four significant figures.

SOLUTION 13-7

In order to solve this problem, we must be sure we know what we're trying to get! Suppose we call our solution x. In the first case, we must solve the following equation for x:

$$e^x = 1,000,000$$

Taking the natural logarithm of each side, we get this:

$$\ln(e^x) = \ln 1,000,000$$

This simplifies to a matter of finding a natural logarithm with a calculator, as follows:

$$x = \ln 1,000,000$$
$$= 13.82$$

In the second case, we must solve the following equation for x:

$$e^x = 0.0001$$

Taking the natural logarithm of each side, we obtain:

$$\ln(e^x) = \ln 0.0001$$

This simplifies, as in the first case, to a matter of finding a natural logarithm with a calculator, as follows:

$$x = \ln 0.0001$$
$$= -9.210$$

PROBLEM 13-8

What is meant by an *order of magnitude* for a positive real-number base b? Give two examples.

SOLUTION 13-8

Let b be a positive real constant. Consider the following two exponentials:

$$p = b^x$$

$$q = b^{(x+n)}$$

where x is a real number and n is a positive integer. In this case, q is n orders of magnitude greater than p in base b.

To demonstrate how this works by an example, take the facts that $10^{-2} = 0.01$ and $10^3 = 1000$. This means that in the base 10, the number 1000 is five orders of magnitude greater than the number 0.01, because the power of 10 for 1000 is five greater than the power of 10 for 0.01. That is, $3 - (-2) = 5$.

Another example involves powers of 2. An increase of one order of magnitude in base 2 is the equivalent of multiplying a given number by 2 (doubling it). Thus, starting with $2^2 = 4$:

$$2^2 = 4$$

$$2^3 = 8 = 2 \times (2^2)$$

$$2^4 = 16 = 2 \times (2^3)$$

$$2^5 = 32 = 2 \times (2^4)$$

$$2^6 = 64 = 2 \times (2^5)$$

and so on. Conversely, a decrease of one order of magnitude in base 2 is the equivalent of dividing a given number by 2 (halving it). Thus, again starting with $2^2 = 4$:

$$2^2 = 4$$

$$2^1 = 2 = 1/2 \times (2^2)$$

$$2^0 = 1 = 1/2 \times (2^1)$$

$$2^{-1} = 1/2 = 1/2 \times (2^0)$$

$$2^{-2} = 1/4 = 1/2 \times (2^{-1})$$

and so on. Orders of magnitude in the base 2 are unique because of their repetitive doubling and halving properties. This makes them useful in digital electronics and computing applications.

Quick Practice

Here are some practice problems that cover the material presented in this chapter. Solutions follow the problems.

PROBLEMS

1. Consider the two numbers $x = 2.3713018568$ and $y = 0.902780337$. Find the product xy, using common logarithms, to four significant figures. (Ignore, for the moment, the fact that a calculator can easily be used to solve this problem without using logarithms in any form!)

2. Approximate the product of the two numbers xy from Problem 1, but use natural logarithms instead. Show that the result is the same. Express the answer to four significant figures.

3. The *power gain* of an electronic circuit, in units called decibels (dB), is calculated according to the following formula:

$$\text{Gain (dB)} = 10 \log (P_{out}/P_{in})$$

where P_{out} is the output signal power and P_{in} is the input signal power, both specified in watts. Suppose the audio input to the left channel of a high-fidelity amplifier is 0.535 watts, and the output is 23.7 watts. What is the power gain of this circuit in decibels? Round off the answer to three significant figures.

4. Suppose the audio output signal in the scenario of Problem 3 is run through a long length of speaker wire, so that instead of the 23.7 watts that appears at the left-channel amplifier output, the speaker only gets 19.3 watts. What is the power gain of the length of speaker wire, in decibels? Round off the answer to three significant figures.

5. If a positive real number increases by a factor of exactly 10, how does its common (base-10) logarithm change?

SOLUTIONS

1. Use the property of common logarithms that converts a product into a sum. That's the following formula:

$$\log xy = \log x + \log y$$

In this case, $x = 2.3713018568$ and $y = 0.902780337$. Using a scientific calculator, we get:

$$\log (2.3713018568 \times 0.902780337)$$
$$= \log 2.3713018568 + \log 0.902780337$$
$$= 0.37498684137 + (-0.0444179086)$$
$$= 0.37498684137 - 0.0444179086$$
$$= 0.3305689328$$

This is the common logarithm of the product we wish to find. If we find the common inverse logarithm of this, we'll get the desired result. Inputting this to a calculator and then rounding to four significant figures:

$$\log^{-1} (0.3305689328) = 2.141$$

2. Use the property of natural logarithms that converts a product into a sum. That's the following formula:

$$\ln xy = \ln x + \ln y$$

In this case, $x = 2.3713018568$ and $y = 0.902780337$. Therefore:

$$\ln (2.3713018568 \times 0.902780337)$$
$$= \ln 2.3713018568 + \ln 0.902780337$$
$$= 0.86343911100 + (-0.102276014)$$
$$= 0.86343911100 - 0.102276014$$
$$= 0.761163097$$

This is the natural logarithm of the product we wish to find. If we find the natural inverse logarithm of this, we'll get the desired result. Inputting this to a calculator and then rounding to four significant figures:

$$\ln^{-1} (0.761163097) = 2.141$$

3. In this scenario, $P_{out} = 23.7$ and $P_{in} = 0.535$. Plug these numbers into the formula for gain in decibels, and then round off as follows:

$$\begin{aligned} \text{Gain (dB)} &= 10 \log (P_{out}/P_{in}) \\ &= 10 \log (23.7/0.535) \\ &= 10 \log 44.299 \\ &= 10 \times 1.6464 \\ &= 16.5 \text{ dB} \end{aligned}$$

4. In this scenario, $P_{out} = 19.3$ and $P_{in} = 23.7$. Plug these numbers into the formula for gain in decibels, and then round off as follows:

$$\begin{aligned} \text{Gain (dB)} &= 10 \log (P_{out}/P_{in}) \\ &= 10 \log (19.3/23.7) \\ &= 10 \log (0.81435) \\ &= 10 \times (-0.089189) \\ &= -0.892 \text{ dB} \end{aligned}$$

5. If a positive real number increases by a factor of 10, then its common logarithm increases (it becomes more positive or less negative) by 1. This is true regardless of the value of the original number, as long as it is positive.

Quiz

This is an "open book" quiz. You may refer to the text in this chapter. A good score is 8 correct. Answers are in the back of the book.

1. Refer to Fig. 13-9. Assume the numerical values indicated are exact. What is the equation that line P represents?
    (a) $y = \ln (2x)$.
    (b) $y = \ln (-x)$.
    (c) $y = 0.5 \ln x$.
    (d) None of the above.

2. Refer to Fig. 13-9. Assume the numerical values indicated are exact. What is the equation that line Q represents?

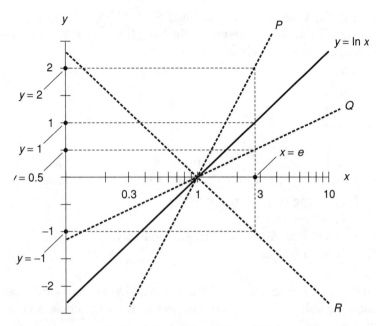

Fig. 13-9. Illustration for Quiz Questions 1 through 3. In this illustration, the y axis intersects the x axis at the point where $x = 0.1$.

(a) $y = \ln (2x)$.
(b) $y = \ln (-x)$.
(c) $y = 0.5 \ln x$.
(d) None of the above.

3. Refer to Fig. 13-9. Assume the numerical values indicated are exact. What is the equation that line R represents?

(a) $y = \ln (2x)$.
(b) $y = \ln (-x)$.
(c) $y = 0.5 \ln x$.
(d) None of the above.

4. What is the common antilogarithm of 0.0036760, accurate to five significant figures? You may use a calculator.

(a) −5.6059.
(b) −2.4346.
(c) 1.0037.
(d) 1.0085.

5. What is the value of $10^{4.553}$ divided by $10^{3.553}$? Express the answer to four significant figures, assuming the base 10 is exact. You may use a calculator, although you shouldn't need one.

   (a) 100.0.
   (b) 10.00.
   (c) 1.000.
   (d) 0.1000.

6. Suppose you are given the equation $e^x = 2.54 \times 10^{-231}$. What can you say about the value of x, without actually solving the equation?

   (a) It is a large positive real number.
   (b) It is between 0 and 1.
   (c) It is a real number and is large negatively.
   (d) It is not a real number.

7. What is ln 10 to seven significant figures? Do not use a calculator. You should be able to infer the correct answer from the choices below without using one.

   (a) 1.000000.
   (b) 2.302585.
   (c) 2.202647×10^4.
   (d) It is not defined.

8. What is ln (-10) to seven significant figures? Do not use a calculator. You should be able to infer the correct answer from the choices below without using one.

   (a) -1.000000.
   (b) -2.302585.
   (c) -2.202647×10^4.
   (d) It is not defined.

9. The range of the common logarithm function (as opposed to its domain) extends over the set of

   (a) nonzero real numbers.
   (b) positive real numbers.
   (c) negative real numbers.
   (d) all real numbers.

10. Let x be a real number that begins at 0 and then becomes larger and larger negatively without limit. What happens to the value of e^x?

 (a) It starts out at 0, and then grows larger and larger positively without limit.

 (b) It starts out at 1, and then passes through 0 and grows larger and larger negatively without limit.

 (c) It starts out at 1, and then becomes smaller positively, approaching, but never quite reaching, 0.

 (d) It starts out at 0, and then becomes larger positively, approaching, but never quite reaching, e.

14

Differentiation in One Variable

This chapter is a quick review of the mechanical basics of differentiation (or *differential calculus*). It is assumed that you have already had some first-year differential calculus. For a more theoretical treatment, *Calculus Demystified* by Steven G. Krantz (McGraw-Hill, New York, 2003) is recommended.

Definition of the Derivative

The *derivative* is a special way of treating a mathematical function in order to get another function. The most common application of derivatives in physics and engineering is the determination of the rate at which a quantity changes at specific points in time. This is known as *instantaneous rate of change*.

WHAT IS A FUNCTION?

Before we formally define the derivative, we should clarify what is meant by the term *function*. You have seen this term in this book before, but now it's time to present a solid definition of it. In calculus, it's vital to know exactly what constitutes a mathematical function—and what does not.

A function is a special sort of mathematical *relation* between two or more variables. A relation defines, by means of one or more equations, how two or more quantities compare with each other as their values vary. A function takes a more active role. It transforms, or *maps*, a quantity or quantities represented by one or more independent variables into a quantity or quantities represented by one or more dependent variables.

The simplest functions are those that have one independent variable and one dependent variable. (Independent and dependent variables were introduced in Chapter 4.) These are known as *single-variable functions* or *functions in one variable*, because they have only one independent variable on which the function operates. Single-variable functions are the only types of functions we will deal with here.

In a single-variable function, changes in the value of the independent variable can, in some situations, be envisioned as causative factors in the variations of the value of the dependent variable. Consider the following real-world statements.

- The outdoor air temperature is a function of the time of day.
- The sunrise time on June 21 is a function of the latitude of the observer.
- The time required for a wet rag to dry is a function of the air temperature.

In scientific and engineering applications, we must be careful when equating the notion of a mathematical function with the notion of a function in nature, human behavior, or any other scenario that isn't purely mathematical. In particular, we must avoid the temptation to suppose that functions and causative relations are always the same thing.

We write $f(x) = y$ or $y = f(x)$ to indicate that a function f maps values of x, the independent variable, to values of y, the dependent variable. We can also say that the function f assigns values of y to values of x.

WHAT IS NOT A FUNCTION?

A relation can be a function only when every element in the set of its independent variables has at most one corresponding element in the set of dependent variables. If a given value of the dependent variable has more than one independent-variable value corresponding to it, then that relation might be a function. But if a given value of the independent variable has two or more corresponding values for the dependent variable, then the relation is not a function.

When graphed, the line or curve representing a function never has more than one point that intersects a "movable straight line" that is always parallel to the dependent-variable axis. Examples of functions of the form $y = f(x)$ include:

$$y = f(x) = 2x^2 + 3x + 4$$

$$y = f(x) = \sin x$$

$$y = f(x) = \log x$$

$$y = f(x) = e^x$$

Here are some examples of relations, where y is the dependent variable, that aren't functions of the independent variable x:

$$x^2 + y^2 = 4$$

$$x = y^2$$

$$x = \sin y$$

DOMAIN AND RANGE

Let's also make sure we know exactly what we're talking about when we mention the *domain* and the *range* of a function! These terms have been used loosely, but now it's time to get more rigorous about their meanings.

Let f be a function whose independent variable is x and whose dependent variable is y. Let X be the set of all x for which f produces some y such that $f(x) = y$. Then X is called the *domain* of f. Stated in a more simplistic way, the domain of a function is the set of all independent variable values for which that function is defined.

Let g be a function whose independent variable is x and whose dependent variable is y. Let Y be the set of all y for which there exists some x such that $g(x) = y$. Then Y is called the *range* of g. Stated in a more simplistic way, the range of a function is the set of all dependent variable values for which that function is defined.

FIRST DERIVATIVE

Let f be a real-number function, let x_0 be an element of the domain of f, and let y_0 be an element of the range of f such that $y_0 = f(x_0)$. Suppose that f is a *continuous function* (meaning that its graph is an unbroken line or curve) in the vicinity of (x_0, y_0) as shown in Fig. 14-1. Let Δx represent a small and shrinking change in x, and let Δy represent the change in y that occurs as a result of Δx. Imagine a movable point (x, y) that is always on the graph of f, and that is always near (x_0, y_0).

The *first derivative* of the function f at the point (x_0, y_0), written $f'(x_0)$, is defined as the limit, as x approaches x_0, of the ratio $\Delta y / \Delta x$. This is equivalent to the limit, as Δx approaches 0, of $\Delta y / \Delta x$. Symbolically this is written as follows:

$$f'(x_0) = \text{Lim}_{\Delta x \to 0} (\Delta y / \Delta x)$$

The word "limit" is abbreviated as "Lim" in equations. The word "approaches" is symbolized by a single-shafted arrow pointing to the right.

In Fig. 14-1, the slope of the line L connecting the two points (x_0, y_0) and (x, y) approaches $f'(x_0)$ as x approaches x_0 from either direction (from the left or from the right). For this reason, the derivative $f'(x_0)$ is graphically described as the slope of a *line tangent to the curve* at the point (x_0, y_0).

If f is continuous (its graph is an unbroken line or curve) at all values of x in its domain, then the first derivative of f is defined in general. This can be denoted in several ways. The most common notations are $f'(x)$, $d/dx(f)$, df/dx, and dy/dx.

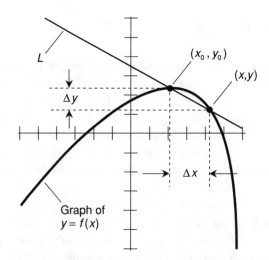

Fig. 14-1. The derivative of a continuous function at a point can be represented by the slope of the curve at that point.

SECOND DERIVATIVE

The *second derivative* of a function f is the derivative of its derivative. This represents the instantaneous rate of change, or slope, of the first derivative function. If $y = f(x)$, this can be denoted in various ways, as follows:

$$f''(x)$$

$$[f'(x)]'$$

$$d^2/dx^2(f)$$

$$d^2f/dx^2$$

$$d^2y/dx^2$$

HIGHER-ORDER DERIVATIVES

The *nth derivative* of a function f is the derivative taken in succession n times, where n is a positive integer. This can be denoted in various ways, as follows:

$$f^{(n)}(x)$$

$$d^n/dx^n(f)$$

$$d^nf/dx^n$$

$$d^ny/dx^n$$

A WARNING ABOUT NOTATION

The superscript n, written all by itself in parentheses after a function designator such as f, denotes the nth derivative, not the nth power. The same is true for superscripts following the letter d or a variable in an expression such as d^ny/dx^n, which represents the nth derivative of y with respect to x. Be careful when encountering expressions in calculus where superscripts are used! Be sure you know whether the superscript refers to a derivative, or is actually intended to be an exponent. This can sometimes be unclear if it is not explained in the text that accompanies mathematical equations involving derivatives.

PROBLEM 14-1

The graph of a circle is not a function in Cartesian coordinates (the xy-plane), regardless of whether x or y is considered the independent variable. This is because there are some values of x for which there exist more than one corresponding value of y. How can a circle be rendered as a true function?

SOLUTION 14-1

Consider the polar coordinate plane represented by ordered pairs (θ, r), in which the angle θ is the independent variable and the radius r is the dependent variable. In this system, you can use a "radial ray test" rather than the "straight line test" you use in the Cartesian plane. Imagine a rotating ray M with its endpoint at the origin. A complete 360° rotation of the ray covers all possible values of θ, the independent variable. A relation defined as $r = f(\theta)$ is a function if and only if every value of θ maps into at most one corresponding value of r. Any circle centered at the origin of such a system has an equation represented as $r = f(\theta) = k$, where k is a positive real-number constant. As you can verify by drawing a simple illustration, any such equation represents a function in polar coordinates. It is the analog of the straight line $y = f(x) = k$, where k is a real-number constant, in Cartesian coordinates.

PROBLEM 14-2

Consider the circle with the following equation in Cartesian coordinates:

$$x^2 + y^2 = 4$$

How can this equation be modified so it becomes a function of x in Cartesian coordinates?

SOLUTION 14-2

Most relations that are not functions can be modified so the result is a function. This is done by restricting one, or both, of the variables to only those values for which there exists at most one value of the dependent variable. In the case of the above equation, which represents a circle centered at the origin with a radius of 2, imagine restricting the values of y so that they are positive (that is, $y > 0$). When this is done, the result is the equivalent of "cutting off" the lower half of the circle. This produces a function whose equation can be derived from the original equation like this:

$$x^2 + y^2 = 4$$

$$y^2 = 4 - x^2$$

$$y = (4 - x^2)^{1/2}$$

$$f(x) = (4 - x^2)^{1/2}$$

The 1/2 power represents the positive square root. If a radical symbol is used over the quantity $(4 - x^2)$ to represent its square root, then a plus sign must be placed in front of the radical symbol to indicate that only the positive square-root values are to be considered.

PROBLEM 14-3
In what other ways can the aforementioned equation be modified in order to get a function of x in Cartesian coordinates?

SOLUTION 14-3
Any restriction that prevents the mapping of an independent-variable value to more than one dependent-variable value will work. As a supplemental exercise, you can draw illustrations to show several such situations.

Properties of Derivatives

Normally, derivatives are defined only for functions, and not for relations that are not functions. This prevents ambiguity that could otherwise occur. Derivatives have some properties that are worth memorizing.

DERIVATIVE OF A CONSTANT FUNCTION

The derivative of a *constant function* is always the *zero function* (a function whose value is equal to 0 throughout its domain). Let f be a function of x such that $f(x) = k$, where k is a real-number constant. Then the following equation always holds:

$$d(f)/dx = 0$$

DERIVATIVE OF SUM OF FUNCTIONS

Let f and g be two different functions of x. Suppose that $f + g = f(x) + g(x)$ for all x in the domains of both f and g. Then:

$$d(f + g)/dx = df/dx + dg/dx$$

That is, the derivative of a sum is equal to the sum of the derivatives. This can be extrapolated to a sum of three, four, or any number of derivatives.

DERIVATIVE OF DIFFERENCE OF TWO FUNCTIONS

Let f and g be two different functions of x. Suppose that $f - g = f(x) - g(x)$ for all x in the domains of both f and g. Then:

$$d(f - g)/dx = df/dx - dg/dx$$

DERIVATIVE OF FUNCTION MULTIPLIED BY A CONSTANT

Let f be a function of x, and let k be a constant. Then the following holds true:

$$d(kf)/dx = k(df/dx)$$

DERIVATIVE OF PRODUCT OF TWO FUNCTIONS

Let f and g be two different functions of x. Define the product of f and g as follows:

$$f \times g = f(x) \times g(x)$$

for all x in the domains of both f and g. Then:

$$d(f \times g)/dx = [f(x) \times (dg/dx)] + [g(x) \times (df/dx)]$$

The square brackets, while technically not necessary, are added for clarity.

DERIVATIVE OF PRODUCT OF THREE FUNCTIONS

Let f, g, and h be three different functions of x. Define the product of f, g, and h as follows:

$$f \times g \times h = f(x) \times g(x) \times h(x)$$

for all x in the domains of f, g, and h. Then:

$$d(f \times g \times h)/dx = [f(x) \times g(x) \times (dh/dx)] \\ + [f(x) \times h(x) \times (dg/dx)] + [g(x) \times h(x) \times (df/dx)]$$

The square brackets, while technically not necessary, are added for clarity.

DERIVATIVE OF QUOTIENT OF TWO FUNCTIONS

Let f and g be two different functions, and define $f/g = [f(x)]/[g(x)]$ for all x in the domains of both f and g. Then:

$$d(f/g)/dx = \{[g(x) \times (df/dx)] \times [f(x) \times (dg/dx)]\}/g^2$$

where $g^2 = g(x) \times g(x)$, not to be confused with the second derivative d^2g/dg^2. The extra brackets in the numerator are added for clarity.

RECIPROCAL DERIVATIVES

Let f be a function, and let x and y be variables such that $y = f(x)$. The following formulas hold:

$$dy/dx = 1/(dx/dy)$$

$$dx/dy = 1/(dy/dx)$$

DERIVATIVE OF A VARIABLE RAISED TO A POWER, TIMES A CONSTANT

Let f be a function, let x be a variable. Suppose f is of the following form:

$$f(x) = kx^n$$

where k is a real-number constant and n is an element of the set $N = \{0, 1, 2, 3,\ldots\}$. Then the first derivative of f is given by the following general equation:

$$f'(x) = nkx^{(n-1)}$$

CHAIN RULE

Let f and g be two functions. Suppose g is a function of x. Then the derivative of $f[g(x)]$ is given by the following formula:

$$f[g(x)]' = f'[g(x)] \times g'(x)$$

PROBLEM 14-4

Consider the following three functions of a variable x:

$$f(x) = 3x^2$$

$$g(x) = 7x$$

$$h(x) = 6$$

Now consider the function $k(x) = 3x^2 + 7x + 6$. What is its derivative?

SOLUTION 14-4

In order to solve this, we will need to take advantage of the rules for the derivatives of:

- a variable raised to a power and then multiplied by a constant
- a constant
- a sum of functions

First, note the following:

$$f'(x) = (2 \times 3)x^{2-1} = 6x^1 = 6x$$

$$g'(x) = (7 \times 1)x^{1-1} = 7x^0 = 7$$

$$h'(x) = 0$$

The function $k(x)$ is the sum of the original three functions. Therefore, its derivative is the sum of the derivatives of the original three functions:

$$k'(x) = f'(x) + g'(x) + h'(x)$$
$$= 6x + 7$$

PROBLEM 14-5

Consider the three functions $f(x)$, $g(x)$, and $h(x)$ defined in the previous problem. Now consider the following function:

$$p(x) = f(x) \times g(x) \times h(x)$$

What is the derivative of this function?

SOLUTION 14-5

Use the formula for the derivative of the product of three functions:

$$d(f \times g \times h)/dx = [f(x) \times g(x) \times (dh/dx)] + [f(x) \times h(x) \times (dg/dx)]$$
$$+ [g(x) \times h(x) \times (df/dx)]$$
$$= (3x^2 \times 7x \times 0) + (3x^2 \times 6 \times 7) + (7x \times 6 \times 6x)$$
$$= 0 + 126x^2 + 252x^2$$
$$= 378x^2$$

PROBLEM 14-6

Solve the previous problem in a less messy way.

SOLUTION 14-6

Note that $p(x)$ is a product of three expressions:

$$p(x) = (3x^2)(7x)(6)$$

These expressions can be multiplied together, getting:

$$p(x) = 126x^3$$

The derivative can be found according to the rule for the derivative of a variable raised to a power and then multiplied by a constant:

$$p'(x) = (126 \times 3)x^{3-1} = 378x^2$$

Properties of Curves

Derivatives are useful for defining the characteristics of graphed functions in the Cartesian plane, especially if those graphs are continuous curves. Here are some of the most significant principles of derivatives involving curves.

LINE TANGENT TO CURVE AT POINT (x_0, y_0)

Let f be a continuous function such that $y = f(x)$. Let (x_0, y_0) be a point on the graph of f, and suppose f is continuous at (x_0, y_0). Let L be a line tangent to the graph of f, and suppose L passes through (x_0, y_0) as shown in Fig. 14-2. Suppose the derivative of f at (x_0, y_0) is equal to some real number m. The equation of line L is given by:

$$y - y_0 = m(x - x_0)$$

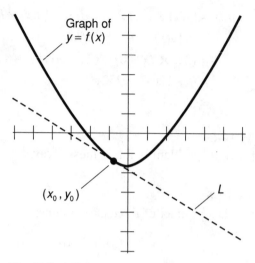

Fig. 14-2. A line tangent to a curve at a point.

If the derivative of f at (x_0, y_0) is equal to 0, then the equation of the line L tangent to f at that point is given by:

$$y = y_0$$

LINE NORMAL TO CURVE AT POINT (x_0, y_0)

Let f be a continuous function such that $y = f(x)$. Let (x_0, y_0) be a point on the graph of f, and suppose f is continuous at (x_0, y_0). Let L be a line normal (perpendicular) to the graph of f, and suppose L passes through (x_0, y_0) as shown in Fig. 14-3. Suppose the derivative of f at (x_0, y_0) is equal to some nonzero real number m. Then the equation of line L is given by the following:

$$y - y_0 = (-x + x_0)/m$$

If the derivative of f at (x_0, y_0) is equal to 0, then the equation of the line L normal to f at that point is given by:

$$x = x_0$$

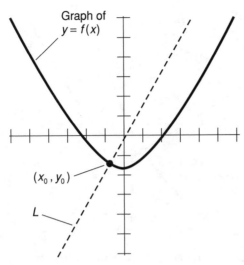

Fig. 14-3. A line normal to a curve at a point.

ANGLE OF INTERSECTION BETWEEN CURVES

Let f and g be functions such that $y = f(x)$ and $y = g(x)$, and suppose that both f and g are continuous in the vicinity of a point (x_0,y_0) at which the graphs of the functions intersect, as shown in Fig. 14-4. Suppose the derivative of f at (x_0,y_0) is equal to some nonzero real number m, and the derivative of g at (x_0,y_0) is equal to some nonzero real number n. Then the acute angle θ at which the graphs intersect is given by the following:

$$\theta = \tan^{-1}\left[(m - n)/(mn + 1)\right]$$

if $m > n$, and

$$\theta = 180° - \tan^{-1}[(m - n)/(mn + 1)]$$

if $m < n$, for angle measures in degrees. If $m = n$, then $\theta = 0°$ or else $\theta = 180°$, and the two curves are tangent to each other at (x_0,y_0). For angle measures in radians, substitute π for $180°$.

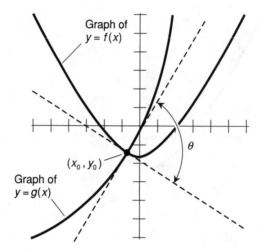

Fig. 14-4. The angle of intersection between two curves at a point.

LOCAL MAXIMUM

Let f be a function such that $y = f(x)$. Let (x_0, y_0) be a point on the graph of f, and suppose f is continuous at (x_0, y_0). Also, suppose the following are both true:

$$f'(x_0) = 0$$
$$f''(x_0) < 0$$

Then (x_0, y_0) is a *local maximum* in the graph of f. An example is shown in Fig. 14-5.

LOCAL MINIMUM

Let f be a function such that $y = f(x)$. Let (x_0, y_0) be a point on the graph of f, and suppose f is continuous at (x_0, y_0). Also, suppose the following are both true:

$$f'(x_0) = 0$$
$$f''(x_0) > 0$$

Then (x_0, y_0) is a *local minimum* in the graph of f. An example is shown in Fig. 14-6.

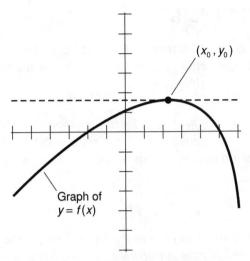

Fig. 14-5. A local maximum value of a function.

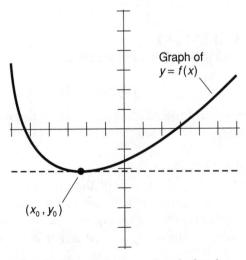

Fig. 14-6. A local minimum value of a function.

INFLECTION POINT

Let f be a function such that $y = f(x)$. Let (x_0, y_0) be a point on the graph of f, and suppose f is continuous at (x_0, y_0). Also, suppose the following is true:

$$f''(x_0) = 0$$

Then (x_0, y_0) is an *inflection point* in the graph of f. Examples are shown in Fig. 14-7. At A, as you move from left to right in the graph, the curve changes from *concave upward* to *concave downward*, and the following inequalities hold:

$$f''(x) > 0 \text{ for } x < x_0$$
$$f''(x) < 0 \text{ for } x > x_0$$

At B, as you move from left to right in the graph, the curve changes from concave downward to concave upward, and the following inequalities hold:

$$f''(x) < 0 \text{ for } x < x_0$$
$$f''(x) > 0 \text{ for } x > x_0$$

The slopes of the tangent lines at the inflection points in both of these examples happens to be 0; the tangent lines appear "horizontal." This is because, in both of these cases, the first derivative happens to be equal to 0 at (x_0, y_0). That does not necessarily have to be true. An inflection point can exist in a curve even when the first derivative is not equal to 0 at that point. The key concept is that the *sense of concavity* reverses at an inflection point.

PROBLEM 14-7
Consider the following function:

$$f(x) = x^2 - 4x + 13$$

The graph of this function is a parabola that opens upward. Thus, it has a point (x_0, y_0) at which the value of $f(x)$ is minimum. What is this point?

SOLUTION 14-7
Find the first derivative of the function, $f'(x)$. The minimum is the point at which the value of this first derivative is equal to 0. Recall that the derivative of a sum is equal to the sum of the derivatives. Therefore, according to the rules outlined earlier in this chapter:

$$f'(x) = 2x - 4$$

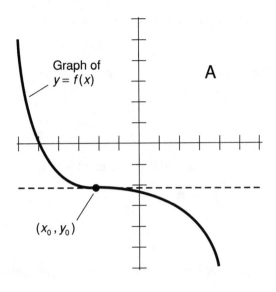

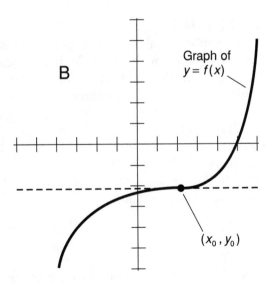

Fig. 14-7. Two examples of inflection points. At A, the function generally decreases; at B, the function generally increases. The tangent lines at the inflection points shown here both happen to have slope 0, but this is a mere coincidence.

Now solve the following equation for x_0:

$$2x_0 - 4 = 0$$

This is a matter of simple algebra:

$$2x_0 - 4 = 0$$
$$2x_0 = 4$$
$$x_0 = 2$$

Now that we know x_0, we can "plug it into" the original function to obtain y_0:

$$y_0 = 2^2 - (4 \times 2) + 13$$
$$= 4 - 8 + 13$$
$$= 9$$

The point at which the graph of $y = f(x)$ attains its minimum is therefore $(x_0, y_0) = (2,9)$.

PROBLEM 14-8
Suppose you're skeptical of the claim, in the statement and solution of the previous problem, that the point in question is a minimum, and not a maximum (which would also produce a derivative equal to 0). Show that the point actually is a minimum.

SOLUTION 14-8
In order to verify this, we must find $f''(x)$, and then determine its value at the point where $x = 2$. The point $(2,9)$ is a minimum if and only if $f''(2) > 0$. The second derivative of a function is equal to the derivative of the first derivative. Thus:

$$f'(x) = 2x - 4$$
$$f''(x) = 2$$

This is the constant function, whose value is always equal to 2. That means $f''(2) = 2$, and therefore, $f''(2) > 0$. This verifies that the point $(2,9)$ is a minimum and not a maximum.

PROBLEM 14-9
Consider the following function:

$$f(x) = -4x^3 + 4x + 5$$

Suppose you are told and assured (so you don't have to prove it) that this function contains an inflection point. Find that inflection point.

SOLUTION 14-9

An inflection point occurs when the second derivative of a function is equal to 0 at that point. The first derivative of f is easy to find, using the rules outlined earlier in this chapter:

$$f'(x) = -12x^2 + 4$$

Next, find the derivative of that:

$$f''(x) = [f'(x)]' = -24x$$

Next, solve the following equation for x_0:

$$-24x_0 = 0$$

We can tell, without doing any algebra, that $x_0 = 0$ is the only solution to this equation. Plugging this into the original function, we can obtain y_0:

$$y_0 = (-4 \times 0^3) + (4 \times 0) + 5$$
$$= 0 + 0 + 5$$
$$= 5$$

The inflection point of the graph of $y = f(x)$ is therefore $(x_0, y_0) = (0,5)$.

Derivatives of Wave Functions

Derivatives are of particular interest in electronics and computer engineering. A circuit called a *differentiator* is used to generate the derivative of a signal wave. The output signal from a differentiator represents the instantaneous rate of change of the input signal. Here are some common periodic wave functions encountered in electronics, and their first derivatives.

DERIVATIVE OF SINE WAVE

The derivative of a *sine wave* (the sine function) is a *cosine wave* (the cosine function). This is the equivalent of a 90° phase shift to the left (Fig. 14-8). The *amplitude*, or signal strength, of the output wave (equivalent to the range of

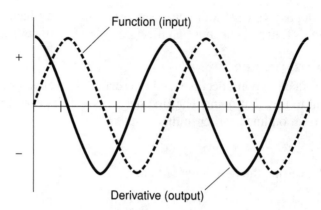

Fig. 14-8. A sine wave (dashed curve) and its derivative as the output of a practical differentiator circuit (solid curve).

the output function, or the *peak-to-peak* value of the wave) depends on the amplitude and frequency of the input sine wave.

DERIVATIVE OF UP-RAMP WAVE

The derivative of an *up-ramp wave* is a positive constant (Fig. 14-9). The value of this constant, which represents the slope of the function, depends on the

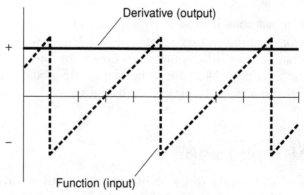

Fig. 14-9. An up-ramp wave (dashed lines) and its derivative as the output of a practical differentiator circuit (solid line).

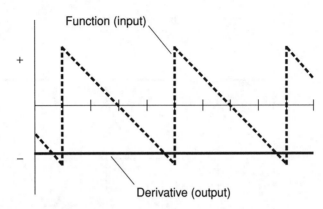

Fig. 14-10. A down-ramp wave (dashed lines) and its derivative as the output of a practical differentiator circuit (solid line).

amplitude of the input wave and also on its frequency. In theory, neither the up-ramp function nor its derivative is defined at the points in time where the input wave is changing state instantaneously (represented by vertical dashed lines). But in practice, the output is continuous because time points have zero duration.

DERIVATIVE OF DOWN-RAMP WAVE

The derivative of a *down-ramp wave* is a negative constant (Fig. 14-10). The value of this constant, which represents the slope of the function, depends on the amplitude of the input wave and also on its frequency. In theory, neither the down-ramp function nor its derivative is defined at the points in time where the input wave is changing state instantaneously (represented by vertical dashed lines). But in practice, the output is continuous because time points have zero duration.

DERIVATIVE OF TRIANGULAR WAVE

The derivative of a *triangular wave* is a *square wave* (Fig. 14-11). The *positive peak amplitude* of the square wave is a positive constant, and the *negative peak amplitude* is a negative constant whose absolute value is the same as that of the positive constant. The output square-wave peak amplitudes depend on the amplitude and frequency of the triangular wave. In theory, the derivative is not defined at the points in time where the input wave is changing from the increas-

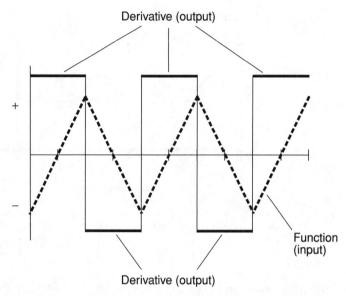

Fig. 14-11. A triangular wave (dashed lines) and its derivative as the output of a practical differentiator circuit (solid lines).

ing condition to the decreasing condition (or vice-versa). These are the points at which the square wave changes state; they have theoretically zero duration. They are represented by vertical solid lines in Fig. 14-11.

DERIVATIVE OF SQUARE WAVE

The derivative of a square wave is zero (Fig. 14-12), representing the absence of a signal. This is true regardless of the amplitude or the frequency of the square wave, because the slope of a square wave function (representing the instantaneous rate of change, and thus the derivative) is equal to 0 at all points for which the function is defined. In theory, neither the square-wave function nor its derivative is defined at the points in time where the input wave is changing state instantaneously (represented by vertical solid lines). But in practice, the output is continuous, because time points have zero duration.

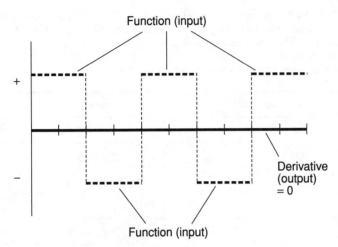

Fig. 14-12.   A square wave (dashed lines) and its derivative as the output of a practical differentiator circuit (solid line).

DERIVATIVE OF COMMON FUNCTIONS

Table 14-1 lists the derivatives of some common mathematical functions.

PROBLEM 14-10
Examine Figs. 14-9 and 14-10, showing the derivatives of up-ramp and down-ramp waves. The outputs in these scenarios are continuous constants in practice, because the transition points (where each ramp ends and the next one begins) have zero duration, and in the "real world" that means they might as well not exist. But when we scrutinize these functions from a rigorous theoretical standpoint, the transition points must be reckoned with! Draw a graph showing the derivatives of the up-ramp and down-ramp functions from Figs. 14-9 and 14-10, but taking the input function transitions into account.

SOLUTION 14-10
Figure 14-13 shows the derivatives of the up-ramp and down-ramp functions (horizontal solid lines), taking the input function transitions

Table 14-1. Derivatives of common functions. The letter a denotes a general real-number constant. The letter n denotes an integer. The letter f denotes a function. The letter x denotes a variable. The letter e represents the exponential constant (approximately 2.71828). A comprehensive table of derivatives can be found online at *www.mathworld.wolfram.com.*

Function	Derivative
$f(x) = a$	$f'(x) = 0$
$f(x) = ax$	$f'(x) = a$
$f(x) = ax^n$	$f'(x) = nax^{n-1}$
$f(x) = 1/x$	$f'(x) = \ln |x|$
$f(x) = \ln x$	$f'(x) = 1/x$
$f(x) = \ln g(x)$	$f'(x) = g^{-1}(x)g'(x)$
$f(x) = 1/x^a$	$f'(x) = -a/(x^{a+1})$
$f(x) = e^x$	$f'(x) = e^x$
$f(x) = a^x$	$f'(x) = a^x \ln a$
$f(x) = a^{g(x)}$	$f'(x) = (a^{g(x)})(\ln a)g'(x)$
$f(x) = e^{ax}$	$f'(x) = ae^x$
$f(x) = e^{g(x)}$	$f'(x) = e^{g(x)} g'(x)$
$f(x) = \sin x$	$f'(x) = \cos x$
$f(x) = \cos x$	$f'(x) = -\sin x$
$f(x) = \tan x$	$f'(x) = \sec^2 x$
$f(x) = \csc x$	$f'(x) = -\csc x \cot x$
$f(x) = \sec x$	$f'(x) = \sec x \tan x$
$f(x) = \cot x$	$f'(x) = -\csc^2 x$
$f(x) = \arcsin x = \sin^{-1}x$	$f'(x) = 1/(1 - x^2)^{1/2}$
$f(x) = \arccos x = \cos^{-1}x$	$f'(x) = -1/(1 - x^2)^{1/2}$
$f(x) = \arctan x = \tan^{-1}x$	$f'(x) = 1/(1 + x^2)$

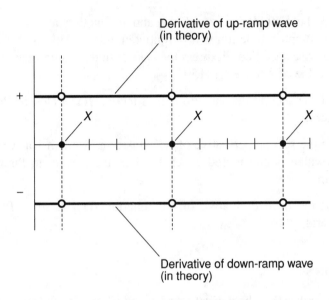

Derivative of up-ramp wave
(in theory)

Derivative of down-ramp wave
(in theory)

X = Input wave transition points

Fig. 14-13.   Illustration for Problem 14-10. Points marked X, and the
vertical dashed lines, represent values for which the input functions
are theoretically undefined.

(points marked X and vertical dashed lines) into account. Theoretically,
the input functions are not defined at the transition points, so the derivatives at those points can't be defined either.

Quick Practice

Here are some practice problems that cover the material presented in this chapter. Solutions follow the problems.

PROBLEMS

1. Find the first derivative of the following function, in general:

$$f(x) = 2x^4 - 6x^2 + \sin x$$

2. Find the first derivative of the above function at $x = \pi/4$. (When a trigonometric function is differentiated, the variable is in radians unless otherwise specified.) Express the answer to three significant figures. Take the value of π to be 3.14159.

3. Find the second derivative of the function $f(x)$ stated in Problem 1, in general.

4. Find the second derivative of $f(x)$ stated in Problem 1 at $x = \pi/4$, where this value is considered exact. Express the answer to three significant figures.

5. Find the third derivative of the function $f(x)$ stated in Problem 1, in general.

SOLUTIONS

1. Remember that the derivative of a sum is equal to the sum of the derivatives. Using this fact, the rule for the derivative of a variable raised to a power and multiplied by a constant, and the fact that the derivative of the sine function is equal to the cosine function, we obtain:

$$f'(x) = (4 \times 2)x^{(4-1)} - (2 \times 6)x^{(2-1)} + \cos x$$
$$= 8x^3 - 12x + \cos x$$

2. Consider $\pi/4 = 3.14159/4 = 0.785398$, and note that $\pi/4$ radians is the equivalent of exactly $45°$ (for use when determining the cosine with a calculator). We obtain the value of the derivative at $x = \pi/4$ this way:

$$f'(\pi/4) = [8 \times (0.785398)^3] - (12 \times 0.785398) + \cos 45°$$
$$= 3.87578 - 9.42478 + 0.70711$$
$$= -4.84189$$

Rounded to three significant figures, this is $f'(\pi/4) = -4.84$.

3. The derivative of the cosine function is equal to the negative of the sine function (from Table 14-1). We find the derivative of $f'(x)$, which is the second derivative of $f(x)$, as follows:

$$f''(x) = (3 \times 8)x^{(3-1)} - (12 \times 1)x^{(1-1)} + (- \sin x)$$
$$= 24x^2 - 12 - \sin x$$
$$= 24x^2 - \sin x - 12$$

The above sum is rearranged in the last line because it is customary to express sums in order by decreasing powers of x.

4. Again consider $\pi/4 = 0.785398$, and again note that $\pi/4$ radians is the equivalent of exactly 45°. The value of the second derivative at $x = \pi/4$ can be found as follows:

$$f''(\pi/4) = [24 \times (0.785398)^2] - \sin 45° - 12$$
$$= 14.8044 - 0.70711 - 12$$
$$= 2.09729$$

Rounded to three significant figures, this is $f''(\pi/4) = 2.10$.

5. The derivative of the sine function is equal to the cosine function (from Table 14-1). We can therefore find the derivative of $f''(x)$, which is the third derivative of $f(x)$, as follows:

$$f'''(x) = (2 \times 24)x^{(2-1)} - \cos x - 0$$
$$= 48x - \cos x$$

Quiz

This is an "open book" quiz. You may refer to the text in this chapter. You may draw diagrams if that will help you visualize things. A good score is 8 correct. Answers are in the back of the book.

1. Consider the equation $x = y^2$. This is a parabola that opens to the right when graphed in the rectangular xy-plane. In this case, y is not a function of x. How can this relation be modified so that y becomes a function of x with a defined, nonempty domain and a defined, nonempty range?

    (a) The allowable values of x can be restricted to the nonnegative real numbers.
    (b) The allowable values of x can be restricted to the negative real numbers.
    (c) The allowable values of y can be restricted to the nonnegative real numbers.
    (d) The allowable values of y can be restricted to real numbers between, but not including, −1 and 1.

2. Which of the following is a rough graph of the modification to the equation $x = y^2$ that results in a function, as defined according to the correct answer to Question 1? In each graph, the horizontal axis represents x, the vertical axis represents y, and a single division on either axis represents an increment of 1 unit.

   (a) Figure 14-14A.
   (b) Figure 14-14B.
   (c) Figure 14-14C.
   (d) None of the above.

3. What is the slope, m, of a line tangent to the graph of the function as defined according to the correct answer to Question 1, at the point where $x = 1$?

   (a) It is not defined.
   (b) $m = 1$.
   (c) $m = 1/2$.
   (d) $m = 2$.

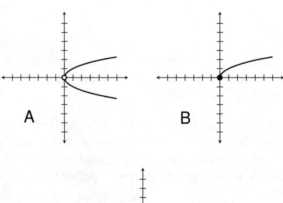

Fig. 14-14.   Illustrations for Quiz Questions 2 through 4.

4. What is the slope, m, of a line tangent to the graph of the function as defined according to the correct answer to Question 1, at the point where $x = 0$?

  (a) It is not defined.
  (b) $m = 0$.
  (c) $m = 1$.
  (d) $m = -1$.

5. Suppose you just bought a new car, and you want to see how fast it can accelerate. You test it on a straight drag strip. A computer is connected to the speedometer, and it generates a speed versus time graph that looks like Fig. 14-15 for speeds ranging from 0 to 25 meters per second. Note that the instantaneous acceleration of an object traveling in a straight line is the derivative of the speed. From Fig. 14-15, it is evident that

  (a) the acceleration is constant.
  (b) the instantaneous acceleration increases as the car goes faster.
  (c) the instantaneous acceleration decreases as the car goes faster.
  (d) the instantaneous acceleration increases at first, and then decreases after the car reaches a certain speed.

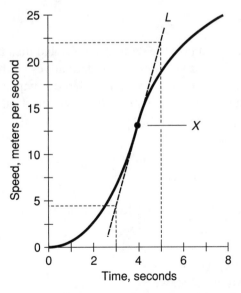

Fig. 14-15. Illustration for Quiz Questions 5 through 7.

6. In the scenario described in Problem 5 and graphed in Fig. 14-15, in what units would acceleration be expressed?

(a) Meters.
(b) Meters per second.
(c) Meters per second per second (or meters per second squared).
(d) Seconds per meter.

7. Consider the situation described in Problems 5 and 6. Line L (the slanted, dashed line) is tangent to the graph of the speed versus time function (the solid curve) at point X. The vertical and horizontal dashed lines intersect line L and the coordinate axes, and are for reference in determining the slope of line L. From this information, it is apparent that the acceleration of the car at point X, expressed in units that arise mathematically when time is defined in seconds and speed is defined in meters per second, is

(a) impossible to figure out without more information.
(b) approximately 22.0/5.0, or 4.4.
(c) approximately 4.2/3.0, or 1.4.
(d) approximately $(22.0 - 4.2)/(5.0 - 3.0)$, or 8.9.

8. Consider a function of the following general form, where x is a real-number variable and n is an integer greater than or equal to 5:

$$f(x) = x^n + x^{n-1} + x^{n-2} + \ldots + x^2 + x + 1$$

Suppose you take the derivative of f, and then take the derivative of f' to obtain f'', and keep on taking derivatives of the derivatives indefinitely. Sooner or later, no matter how large n happens to be, you will end up with a function that is its own derivative. That function will be

(a) a quadratic function.
(b) a linear function.
(c) a constant function.
(d) the zero function.

9. Which of the following functions is its own fourth derivative?

(a) $f(x) = \sin x$
(b) $g(x) = \cos x$
(c) $h(x) = -\sin x$
(d) All of the above

10. In a system of polar coordinates with θ as the independent variable and r as the dependent variable, imagine a function $f(\theta)$ with a derivative as follows:

$$dr/d\theta = f'(\theta) = k$$

where k is a real-number constant. What does the graph of $r = f(\theta)$ look like when the domain is restricted to values between, but not including, 0° and 360°?

(a) A straight ray (half-line).
(b) A half-circle.
(c) A single rotation of a spiral.
(d) Half of a parabola.

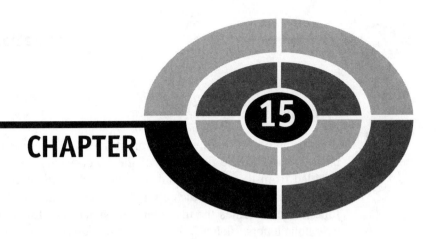

Integration in One Variable

Integration is the opposite of differentiation. *Integral calculus* is used to find areas, volumes, and accumulated quantities. This chapter is a brief review, intended for students who have had some first-year integral calculus. For a theoretical discussion, *Calculus Demystified* by Steven G. Krantz (McGraw-Hill, New York, 2003) is recommended.

What Is Integration?

When a function is differentiated, the result is another function that expresses the instantaneous rate of change of the original function. Integration is, in a sense, the reverse of this. It produces a new function that expresses the cumulative growth of the original function.

THE ANTIDERIVATIVE

Imagine an object moving in a straight line. The function that defines its instantaneous speed is the derivative of the function that defines the distance it has traveled from the starting point (its cumulative displacement). Reversing this, the function that defines the cumulative displacement is the *antiderivative* of the function that defines the instantaneous speed. For functions denoted by lowercase italic letters, such as f, g, or h, the antiderivative is denoted by the uppercase italic counterpart, such as F, G, or H, respectively. The concept of the antiderivative is the basis for mathematical integration.

THE INDEFINITE INTEGRAL

Let f be a continuous real-number function of a variable x. Consider $y = f(x)$. The *indefinite integral* of f is a function F such that $dy/dx = f(x)$, added to a real-number constant c. This is written as follows:

$$\int f(x)\, dx = F(x) + c$$

where dx represents the *differential* of x. The above expression, if read out loud, is "the indefinite integral of $f(x)\ dx$," or "the indefinite integral of $f(x)$ with respect to x."

In this book, we won't get concerned about the precise meaning and significance of the differential. But you should remember that it's customary to include the differential at the end of any integral. Always end the expression with a lowercase italic letter d followed by the variable with respect to which the integration is to be done (x, y, z, θ, ϕ, or whatever).

THE CONSTANT OF INTEGRATION

The constant c, which appears in all indefinite integrals, arises from the fact that the derivative of a constant function is equal to the zero function. Indefinite integration "undoes" differentiation, but this happens in an ambiguous sense. A function has only one specific derivative, but it can have infinitely many indefinite integrals, all of which differ by some value of c, which is known as the *constant of integration*. The antiderivative is a special case of the indefinite integral, where $c = 0$.

AN EXAMPLE

Consider the function $f(x) = x^2$. The derivative of this function is $f'(x) = 2x$. Consider the following indefinite integral:

$$\int 2x \, dx$$

Let's rename the function we are integrating, in order to distract us from the fact that it's something we've already conjured up. Suppose we have this:

$$g(x) = 2x$$

The antiderivative of g is the function G, as follows:

$$G(x) = x^2$$

It's important to note that G is not the only function of x that can be differentiated to get the function g. Any real number c can be added to $G(x)$, and when the resulting function, $G(x) + c$, is differentiated, the result is always equal to $g(x)$. Think of it like this, where c can be any real number whatsoever:

$$G(x) + c = x^2 + c$$

and therefore:

$$[G(x) + c]' = 2x + 0$$
$$= 2x$$
$$= g(x)$$

That means:

$$\int 2x \, dx = x^2 + c$$

THE DEFINITE INTEGRAL

Let f be a continuous real-number function of a variable x. Let a and b be values in the domain of f such that $a < b$. Let F be the antiderivative of f. The *definite integral* of $f(x)$ from $x = a$ to $x = b$ is denoted and defined as follows:

$$\int_a^b f(x) \, dx = F(b) - F(a)$$

The expression to the left of the equals sign is read, "the integral from a to b of $f(x) \, dx$," or "the integral of $f(x) \, dx$ from a to b." The above general rule is known as the *Fundamental Theorem of Calculus*.

When we find a definite integral, the constant of integration subtracts from itself, so it disappears.

The definite integral of a function f can be depicted as the area between the curve and the independent-variable axis (usually the x axis) in the graph of f in rectangular coordinates. This area can be considered for the entire domain of the function f, or for only part of the domain. When a definite integral is considered for only part of the domain, the limiting value or values (a and/or b in the above expression) must be specified. For example, we might consider the definite integral of a function for values of the domain such as:

- All the positive real numbers
- All the negative real numbers
- All the real numbers between 0 and 1
- All the real numbers between $-\pi$ and π

An example of a definite integral is shown in Fig. 15-1. Regions above the x axis are considered to have positive area. Regions below the x axis are considered to have negative area.

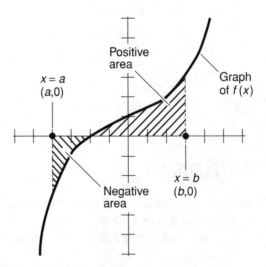

Fig. 15-1.   A definite integral in the *xy*-plane can be represented as the area between the curve and the x axis, as defined between "vertical lines" passing through two points on the x axis. In this case the two points are $(a,0)$ and $(b,0)$. Areas "above" the x axis are considered positive. Areas "below" the x axis are considered negative.

PROBLEM 15-1

Suppose $g(x) = 2x$. Find the definite integral of $g(x)\,dx$ from 0 to 1.

SOLUTION 15-1

In the general expression for the definite integral, let $a = 0$ and let $b = 1$. From the preceding examples, we know that $G(x) = x^2$. Therefore:

$$\int_0^1 g(x)\,dx = G(1) - G(0)$$

Therefore, the value of the definite integral is equal to $1^2 - 0^2 = 1 - 0 = 1$.

PROBLEM 15-2

Again, let $g(x) = 2x$. Find the definite integral of $g(x)\,dx$ from -1 to 1.

SOLUTION 15-2

In the general expression for the definite integral, let $a = -1$ and let $b = 1$. As in the previous problem, $G(x) = x^2$. Therefore:

$$\int_{-1}^1 g(x)\,dx = G(1) - G(-1)$$

In this case, the value of the definite integral is equal to $1^2 - (-1)^2 = 1 - 1 = 0$.

Basic Properties of Integration

Here are several important properties of indefinite integration. These are worth memorizing, because they apply to frequently encountered situations.

INDEFINITE INTEGRAL OF A CONSTANT

Let k be a constant. Let c be the constant of integration. Let x be a variable. The integral of k with respect to x is equal to k times x, plus the constant of integration c. The following formula applies:

$$\int k\,dx = kx + c$$

INDEFINITE INTEGRAL OF A VARIABLE

Let x be a variable. Let c be the constant of integration. The integral of x with respect to itself is equal to $x^2/2$, plus the constant of integration c. The following formula applies:

$$\int x \, dx = x^2/2 + c$$

INDEFINITE INTEGRAL OF A VARIABLE MULTIPLIED BY A CONSTANT

Let x be a variable. Let k be a constant. Let c be the constant of integration. The integral of the product kx with respect to x is equal to k times $x^2/2$, plus the constant of integration c. The following formula applies:

$$\int kx \, dx = kx^2/2 + c$$

INDEFINITE INTEGRAL OF A VARIABLE RAISED TO AN INTEGER POWER

Let x be a variable. Let k be an integer that is not equal to -1. Let c be the constant of integration. The following general rule applies:

$$\int x^k \, dx = [x^{(k+1)}/(k + 1)] + c$$

INDEFINITE INTEGRAL OF A FUNCTION MULTIPLIED BY A CONSTANT

Let f be a function of a variable x. Let k be a constant. The integral of k times $f(x)$ with respect to x is equal to k times the integral of $f(x)$ with respect to x. The following formula applies:

$$\int k \, [f(x)] \, dx = k \int f(x) \, dx$$

INDEFINITE INTEGRAL OF THE SUM OF TWO OR MORE FUNCTIONS

Let $f_1, f_2, f_3, \ldots$, and f_n be functions of a variable x. The integral of the sum of the functions with respect to x is equal to the sum of the integrals of each function with respect to x. The following general rule applies:

$$\int [f_1(x) + f_2(x) + f_3(x) + \ldots + f_n(x)] \, dx$$
$$= \int f_1(x) \, dx + \int f_2(x) \, dx + \int f_3(x) \, dx + \ldots + \int f_n(x) \, dx$$

PROBLEM 15-3

Find the indefinite integral of the function $f(x) = 4x^3 + 3x^2$ with respect to x. That is, determine the following:

$$\int 4x^3 + 3x^2 \, dx$$

SOLUTION 15-3

The function f is the sum of two simpler functions to which the above rules can be easily applied. Consider f to break down into the sum of f_1 and f_2, as follows:

$$f_1(x) = 4x^3$$

$$f_2(x) = 3x^2$$

$$f(x) = f_1(x) + f_2(x) = 4x^3 + 3x^2$$

The antiderivatives, F_1 and F_2, are:

$$F_1(x) = 4x^{(3+1)} / (3 + 1) = x^4$$

$$F_2(x) = 3x^{(2+1)} / (2 + 1) = x^3$$

and therefore:

$$\int 4x^3 + 3x^2 \, dx = x^4 + x^3 + c$$

A Few More Formulas

Here are two important principles, and three more properties, that apply to indefinite integrals. There's also reference to a table of indefinite integrals for commonly encountered functions. The principles of linearity and integration by parts are worth memorizing.

PRINCIPLE OF LINEARITY

Let f and g be defined, continuous functions of x. Let a and b be real-number constants. Then the following general formula applies:

$$\int [a\,f(x) + b\,g(x)]\,dx = a \int f(x)\,dx + b \int g(x)\,dx$$

PRINCIPLE OF INTEGRATION BY PARTS

Let f and g be differentiable, continuous functions of x. Let f' be the derivative of f with respect to x, and let g' be the derivative of g with respect to x. Then the following general formula applies:

$$\int [f(x)\,g'(x)]\,dx = f(x)\,g(x) - \int [f'(x)\,g(x)]\,dx$$

INDEFINITE INTEGRAL OF A RECIPROCAL

Let x be a variable. Let c be the constant of integration. Let ln represent the natural logarithm function. The following rule applies:

$$\int (1/x)\,dx = \ln |x| + c$$

The vertical lines on either side of x represent the absolute value of x. Another way to state this fact is to consider $1/x$ as equal to x raised to the -1 power:

$$\int x^{-1}\,dx = \ln |x| + c$$

INDEFINITE INTEGRAL OF A RECIPROCAL MULTIPLIED BY A CONSTANT

Let x be a variable. Let k be a constant, and let c be the constant of integration. The following general formulas hold:

$$\int (k/x)\, dx = k \ln |x| + c$$

$$\int kx^{-1}\, dx = k \ln |x| + c$$

INDEFINITE INTEGRAL OF A CONSTANT RAISED TO A VARIABLE POWER

Let x be a variable. Let k be a positive real-number constant that is not equal to 1. Let c be the constant of integration. The following general formula holds:

$$\int k^x\, dx = [k^x / (\ln k)] + c$$

INDEFINITE INTEGRALS OF COMMON FUNCTIONS

Table 15-1 lists indefinite integrals of common mathematical functions.

PROBLEM 15-4
Use the principle of linearity to find the following indefinite integral:

$$\int (3 \sin x + 7 \cos x)\, dx$$

SOLUTION 15-4
According to the principle of linearity, the constants can be separated out, and the integral broken down into the sum of two simpler integrals, so we obtain the following:

$$3 \int \sin x\, dx + 7 \int \cos x\, dx$$

Refer to Table 15-1 to obtain the indefinite integrals of the sine and cosine functions. The constants of integration are not identical for the two integrals in this sum, so let's call one of them c_1 and the other one c_2. Then we obtain the following expression:

$$3 (-\cos x + c_1) + 7 (\sin x + c_2) = -3 \cos x + 3c_1 + 7 \sin x + 7c_2$$
$$= 7 \sin x - 3 \cos x + 3c_1 + 7c_2$$

Table 15-1. Indefinite integrals. The letters a and b denote general real-number constants. The letter c denotes the constant of integration. The letter n denotes an integer. The letters f, g, and h denote functions. The letter x denotes a variable. The letter e represents the exponential constant (approximately 2.71828). A comprehensive table of indefinite integrals can be found online at *www.mathworld.wolfram.com*.

Function	Indefinite Integral	Function	Indefinite Integral
$f(x) = a$	$\int f(x)\,dx = a + c$	$f(x) = e^x$	$\int f(x)\,dx = e^x + c$
$f(x) = ax$	$\int f(x)\,dx = (1/2)\,ax^2 + c$	$f(x) = a\,e^{bx}$	$\int f(x)\,dx = (ae^{bx}/b) + c$
$f(x) = ax^2$	$\int f(x)\,dx = (1/3)\,ax^3 + c$	$f(x) = x\,e^{bx}$	$\int f(x)\,dx = b^{-1}x\,e^{bx} - b^{-2}\,e^{bx} + c$
$f(x) = ax^3$	$\int f(x)\,dx = (1/4)\,ax^4 + c$	$f(x) = \ln x$	$\int f(x)\,dx = x \ln x - x + c$
$f(x) = ax^4$	$\int f(x)\,dx = (1/5)\,ax^5 + c$	$f(x) = x \ln x$	$\int f(x)\,dx = (1/2)\,x^2 \ln x - (1/4)\,x^2 + c$
$f(x) = ax^{-1}$	$\int f(x)\,dx = a \ln |x| + c$	$f(x) = \sin x$	$\int f(x)\,dx = -\cos x + c$
$f(x) = ax^{-2}$	$\int f(x)\,dx = -ax^{-1} + c$	$f(x) = \cos x$	$\int f(x)\,dx = \sin x + c$
$f(x) = (ax + b)^{1/2}$	$\int f(x)\,dx = (2/3)\,(ax + b)^{3/2}\,a^{-1} + c$	$f(x) = \tan x$	$\int f(x)\,dx = \ln |\sec x| + c$
$f(x) = (ax + b)^{-1/2}$	$\int f(x)\,dx = 2\,(ax + b)^{1/2}\,a^{-1} + c$	$f(x) = \csc x$	$\int f(x)\,dx = \ln |\tan (x/2)| + c$
$f(x) = (ax + b)^{-1}$	$\int f(x)\,dx = a^{-1}\,[\ln (ax + b)] + c$	$f(x) = \sec x$	$\int f(x)\,dx = \ln |\sec x + \tan x| + c$
$f(x) = (ax + b)^{-2}$	$\int f(x)\,dx = -a^{-1}\,(ax + b)^{-1} + c$	$f(x) = \cot x$	$\int f(x)\,dx = \ln |\sin x| + c$
$f(x) = (ax + b)^n$ where $n \neq -1$	$\int f(x)\,dx = (ax + b)^{n+1}\,(an + a)^{-1} + c$	$f(x) = \arcsin x$	$\int f(x)\,dx = x \arcsin x + (1 - x^2)^{1/2} + c$
$f(x) = ax^n$	$\int f(x)\,dx = ax^{n+1}\,(n + 1)^{-1} + c$ provided that $n \neq -1$	$f(x) = \arccos x$	$\int f(x)\,dx = x \arccos x - (1 - x^2)^{1/2} + c$
$f(x) = a\,g(x)$	$\int f(x)\,dx = a \int g(x)\,dx + c$	$f(x) = \arctan x$	$\int f(x)\,dx = x \arctan x - (1/2) \ln (1 + x^2) + c$
$f(x) = g(x) + h(x)$	$\int f(x)\,dx = \int g(x)\,dx + \int h(x)\,dx + c$		

Let $c = 3c_1 + 7c_2$. Then the original indefinite integral simplifies to this:

$$\int (3 \sin x + 7 \cos x) \, dx = 7 \sin x - 3 \cos x + c$$

 PROBLEM 15-5

Calculate the following indefinite integral:

$$\int (3x)(\cos x) \, dx$$

 SOLUTION 15-5

In order to find the indefinite integral of the product of two functions, integration by parts can be a useful technique. Refer again to the general formula for the principle of integration by parts:

$$\int [f(x) \, g'(x)] \, dx = f(x) \, g(x) - \int [f'(x) \, g(x)] \, dx$$

Let's set $f(x) = 3x$ and $g'(x) = \cos x$. The derivative of $f(x)$ is easy to find:

$$f'(x) = 3$$

Note that $g(x)$ is the antiderivative, or the indefinite integral, of $g'(x)$. Now look at Table 15-1, and note that that $g'(x) = \cos x$. That means the following is true:

$$g(x) = \sin x + c_1$$

where c_1 is a real-number constant, the constant of integration. Now we know the following four facts:

$$f(x) = 3x$$

$$f'(x) = 3$$

$$g(x) = \sin x + c_1$$

$$g'(x) = \cos x$$

Our original integral therefore becomes:

$$\int (3x)(\cos x)\, dx = (3x)(\sin x + c_1) - \int 3\, (\sin x + c_1)\, dx$$

$$= 3x \sin x + 3c_1x - 3 \int (\sin x + c_1)\, dx$$

$$= 3x \sin x + 3c_1x - 3 \left[\int \sin x\, dx + \int c_1\, dx \right]$$

$$= 3x \sin x + 3c_1x - 3\, (-\cos x + c_2 + c_1x + c_3)$$

where c_2 and c_3 are new constants of integration. (The three constants c_1, c_2, and c_3 are not necessarily the same.) Simplifying the above expression is a matter of applying basic rules of algebra:

$$3x \sin x + 3c_1x - 3\, (-\cos x + c_2 + c_1x + c_3)$$

$$= 3x \sin x + 3c_1x - (-3 \cos x + 3c_2 + 3c_1x + 3c_3)$$

$$= 3x \sin x + 3c_1x + 3 \cos x - 3c_2 - 3c_1x - 3c_3$$

$$= 3x \sin x + 3 \cos x - 3c_2 - 3c_3$$

Let's rename the constant $(-3c_2 - 3c_3)$ and call it c. Then we have the solution to the original problem as:

$$\int (3x)(\cos x)\, dx = 3x \sin x + 3 \cos x + c$$

Integrals of Wave Functions

Integrals, like derivatives, are of interest in computer and electronics applications. An *integrator* is a circuit that generates an indefinite integral of an input signal wave. The output signal wave represents the accumulated value of the input signal wave.

SINE VERSUS SINUSOID

A *theoretical sine wave* is represented by the function $f(x) = \sin x$, and no other. However, in electronics practice, all signal waves that have the same geometric shape as this function are called "sine waves," even if they are displaced vertically, displaced horizontally, stretched or squashed vertically, or stretched or squashed horizontally. This can encompass a tremendous variety of signal wave shapes, which engineers call *waveforms*. A technically precise description of such waves is *sinusoid*, a term meaning *sine-like wave*. A sinusoid can be described as a function in the following general form:

$$f(x) = p \sin(qx + r) + c$$

where x is the independent variable expressed in radians, and p, q, r, and s are real-number constants, with the constraint that neither p nor q can be equal to 0. Any function of the following form is also a sinusoid, because the graph of the cosine function has the same general shape as the graph of the sine function:

$$f(x) = p \cos(qx + r) + c$$

The value of p determines the *peak-to-peak amplitude* (the difference between the instantaneous maximum and the instantaneous minimum values of the function). The value of q determines the frequency (the extent to which the wave is horizontally stretched or squashed). The value of r determines the displacement of the wave along the axis of the independent variable, usually to the left or right. The value of c determines the displacement of the wave perpendicular to the axis of the independent variable, usually upward or downward. If the sinusoid is the output of an integrator circuit, c is the practical equivalent of the constant of integration.

INTEGRAL OF THEORETICAL SINE WAVE

The integral of a *theoretical sine wave*, represented by the function $f(x) = \sin x$, is an *inverted cosine wave* (the negative of the cosine function) displaced upward by 1 unit, as shown in Fig. 15-2. In this case, integration begins at the

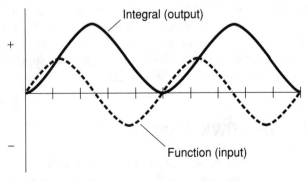

Fig. 15-2.   A sine input wave (dashed curve) and the output of a practical integrator circuit (solid curve).

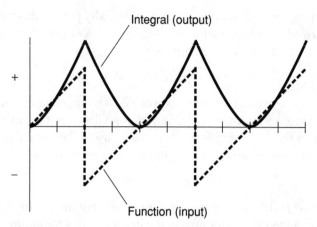

Fig. 15-3.   An up-ramp input wave (dashed lines) and the output of a practical integrator circuit (solid curve).

point where the input wave has an instantaneous amplitude of 0 and is increasing positively. If the integration begins at some other point in the cycle, the upward or downward displacement, which is the practical equivalent of the constant of integration, is some value other than 1.

INTEGRAL OF UP-RAMP WAVE

The integral of an up-ramp wave has the appearance of a series of inverted, vertically displaced half-sine-like waves, as shown in Fig. 15-3. (It is actually a series of parabolic sections.) The direction and extent of the output-wave displacement depends on the starting point of the input wave, where the integration is considered to begin. The amplitude of the output wave depends on the amplitude of the input wave. In theory, the up-ramp function is not defined at the points in time where the input wave is changing state instantaneously (represented by vertical dashed lines). But in practice, the output is defined at these points, because time points have zero duration.

INTEGRAL OF DOWN-RAMP WAVE

The integral of a down-ramp wave has the appearance of a series of vertically displaced half-sine-like waves, as shown in Fig. 15-4. (It is actually a series of parabolic sections.) The direction and extent of the output-wave displacement depends on the starting point of the input wave, where the integration is consid-

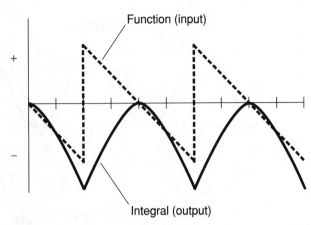

Fig. 15-4.   A down-ramp input wave (dashed lines) and the output of a practical integrator circuit (solid curve)

ered to begin. The amplitude of the output wave depends on the amplitude of the input wave. In theory, the down-ramp function is not defined at the points in time where the input wave is changing state instantaneously (represented by vertical dashed lines). But in practice, the output is defined at these points, because time points have zero duration.

INTEGRAL OF TRIANGULAR WAVE

The integral of a triangular wave is a vertically displaced sine-like wave (not a true sine wave, but a series of parabolic sections, as shown in Fig. 15-5). The direction and extent of the output-wave displacement depends on the starting point of the input wave, where the integration is considered to begin. The amplitude of the output wave depends on the amplitude of the input wave.

INTEGRAL OF SQUARE WAVE

The integral of a square wave is a vertically displaced triangular wave (Fig. 15-6). The direction and extent of the output-wave displacement depends on the starting point of the input wave, where the integration is considered to begin. The amplitude of the output wave depends on the amplitude of the input wave. In theory, the square-wave function is not defined at the points in time where the input wave is changing state instantaneously (represented by vertical dashed lines). But in practice, the output is defined, because time points have zero duration.

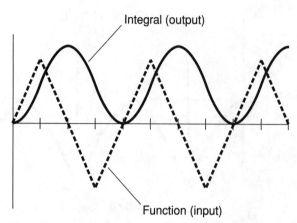

Fig. 15-5. A triangular input wave (dashed lines) and the
output of a practical integrator circuit (solid curve).

PROBLEM 15-6

Suppose that an integrator circuit is supplied with a continuous, positive input voltage. This means that the input function is a positive constant. The indefinite integral of a positive constant function is a straight
line with a positive slope. What does this mean in a practical circuit?
Can the output of a real-life circuit be a positive voltage that increases
without limit?

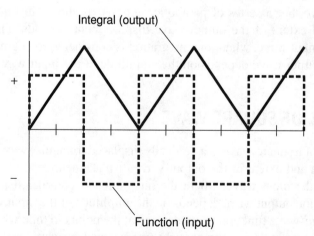

Fig. 15-6. A square input wave (dashed lines) and the output of
a practical integrator circuit (solid lines).

SOLUTION 15-6

If an electronic integrator is supplied with a continuous, positive input voltage, the theoretical output is a positive voltage that starts out at 0 and rises indefinitely at a constant rate, as shown in Fig. 15-7A. But in an actual circuit, the output voltage rises up to a certain limit, as shown at B, and then remains there until the input voltage is removed. In order to bring the output voltage back down to 0, it may be necessary to temporarily disconnect the device from the power source (battery or power supply).

PROBLEM 15-7

How does the shape of the integral of a sinusoid vary when the values of the constants p and q (described on page 349), vary for the input wave? What does vertical displacement of a wave function represent in a real-world electronic system?

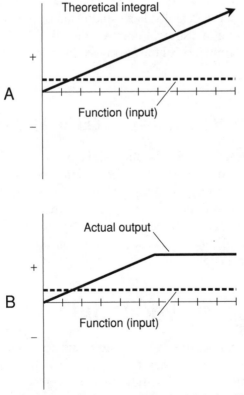

Fig. 15-7. Illustrations for Problem and Solution 15-6. At A, the theoretical situation; at B, a real-life scenario.

SOLUTION 15-7

The integral of a sinusoid is always another sinusoid. The peak-to-peak amplitude of the output sinusoid depends on the peak-to-peak amplitude of the input sinusoid, and also on its frequency. That is, it depends on the values of the constants p and q. The value of p affects the amplitude; the value of q affects the frequency. In electronic circuits, the vertical displacement of the output wave represents a *direct-current* (DC) *component*. This displacement can be as large, either positively or negatively, as half the peak-to-peak amplitude of the input wave. It is the equivalent of the constant of integration.

Examples of Definite Integration

Definite integration can be used to find the area under a curve between two points, as we have already seen. It can also be used to define the relationship among acceleration, speed, and displacement (distance traveled) as functions of time for objects moving in a straight line. Another useful application of definite integration is the determination of the average value of a function over a specific interval.

DISPLACEMENT, SPEED, AND ACCELERATION

In Newtonian physics, the speed function is the derivative of the displacement function with respect to time, and the acceleration function is the derivative of the speed function with respect to time. Conversely, the speed function is the indefinite integral of the acceleration function with respect to time, and the displacement function is the indefinite integral of the speed function with respect to time. This is a simplistic way to state the relationship, but it's worth remembering as a general principle.

DISTANCE AND SPEED VERSUS TIME

Imagine a car that starts out from a "dead stop" and accelerates in a straight line on a level road. Suppose that when the driver steps on the gas pedal, the car exhibits the following speed-versus-time function s, where $s(t)$ is the speed in meters per second and t is the time in seconds:

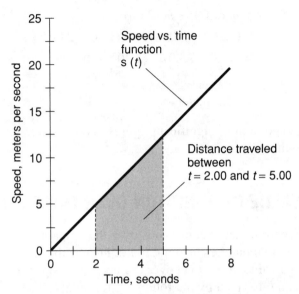

Fig. 15-8.   Determination of distance traveled between two
time points when the function of speed versus time is known.
Illustration for Problems 15-8 and 15-9.

$$s(t) = 2.40t$$

This function is graphed in Fig. 15-8. We can use integration to derive a function of the distance in meters that the car has traveled versus the elapsed time after the start. Let $m(t)$ represent the distance function. (The letter m stands for "meters." We don't want to use d for "distance," because this could get confused with the differential.) Then:

$$m(t) = \int s(t)\, dt$$

$$= \int 2.40t\, dt$$

$$= 1.20t^2 + c$$

where c is the constant of integration.

Suppose we want to find out how far the car goes between $t = 2.00$ and $t = 5.00$. This is the distance traveled between a time point 2.00 seconds after the start, until a time point 5.00 seconds after the start. Let's call this distance z. Then z is equal to $m(t)$ evaluated from $t = 2.00$ to $t = 5.00$, as follows:

$$z = m\,(5.00) - m\,(2.00)$$
$$= (1.20 \times 5.00^2 + c) - (1.20 \times 2.00^2 + c)$$
$$= (1.20 \times 25.0) - (1.20 \times 4.00)$$
$$= 30.0 - 4.80$$
$$= 25.2 \text{ meters}$$

We're allowed to go to three significant figures in our answer, because our input data is given to that level of accuracy.

AVERAGE VALUE OF FUNCTION OVER INTERVAL

Let $f(x)$ be a function that is continuous over the domain from $x = a$ to $x = b$, where a and b are real numbers and $a < b$. Let $F(x)$ be the antiderivative of $f(x)$. Then the average value, A, of $f(x)$ over the open, half-open, or closed interval bounded by a and b is given by the following formula:

$$A = [F(b) - F(a)] / (b - a)$$

This generalized principle is illustrated in Fig. 15-9.

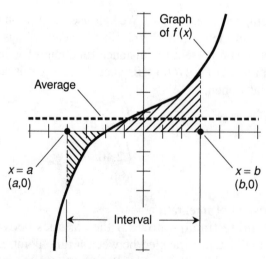

Fig. 15-9. Generalized principle of the average value of a function over an interval.

Geometrically, this is the equivalent of finding the "height" (either positive or negative) of a rectangle whose area is the same as the area between the curve and the *x* axis between "vertical" lines passing through the points $x = a$ and $x = b$. Note that $F(b) - F(a)$ is the same as the definite integral from *a* to *b* of $f(x)$ with respect to *x*. That is the area between the curve and the *x* axis between "vertical" lines passing through the points $x = a$ and $x = b$, and is also the area of the rectangle whose height corresponds to the average value of $f(x)$ over the interval from $x = a$ to $x = b$.

PROBLEM 15-8

Use integration to find the average speed of the car in the example discussed in "Distance and Speed versus Time" above and illustrated in Fig. 15-8, for the interval between the time points $t = 2.00$ and $t = 5.00$.

SOLUTION 15-8

Let $S(t)$ be the antiderivative of $s(t)$ in the above situation. This function is the same as $m\ (t)$ with $c = 0$. That is:

$$S(t) = 1.20t^2$$

If *A* is the average speed over the interval from $t = 2.00$ to $t = 5.00$, then:

$$A = [S\ (5.00) - S\ (2.00)]\ /\ (5.00 - 2.00)$$
$$= [(1.20 \times 5.00^2) - (1.20 \times 2.00^2)]\ /\ (5.00 - 2.00)$$
$$= [(1.20 \times 25.0) - (1.20 \times 4.00)]\ /\ 3.00$$
$$= 25.2\ /\ 3.00$$
$$= 8.40 \text{ meters per second}$$

As in the example in "Distance and Speed versus Time" above, we're allowed to go to three significant figures in our answer, because our input data is given to that level of accuracy.

PROBLEM 15-9

Verify that the answer obtained in Solution 15-8 is correct by finding the arithmetic mean, or average, of the values of the speed function $s\ (t)$ for the time points $t = 2.00$ and $t = 5.00$. (Averaging will work in this situation only because the function of speed versus time is linear. If it were not linear, averaging would not work, and integration would have to be used.)

SOLUTION 15-9

First, find the values of the function $s(t)$ for $t = 2.00$ and $t = 5.00$. These are:

$$s(2.00) = 2.40 \times 2.00$$
$$= 4.80 \text{ meters per second}$$

$$s(5.00) = 2.40 \times 5.00$$
$$= 12.0 \text{ meters per second}$$

The average, A, of these two values is:

$$A = (4.80 + 12.0)/2$$
$$= 16.8/2$$
$$= 8.40 \text{ meters per second}$$

Here, the value of 2 in the denominator is mathematically exact, because the arithmetic mean of two numbers is defined as precisely half the sum of those two numbers. Therefore, we are justified in going to three significant figures for the final answer, because our input data is given to that level of accuracy.

Quick Practice

Here are some practice problems that cover the material presented in this chapter. Solutions follow the problems.

PROBLEMS

1. Find the following indefinite integral:

$$\int (24y^5 + 15y^4 - 9y^2 + 4y - 4)\, dy$$

Assume that the values of the coefficients are exact, so there is no need to be concerned about significant figures.

2. Find the indefinite integral of the result of the above problem. That is, find:

$$\int \left[\int (24y^5 + 15y^4 - 9y^2 + 4y - 4)\, dy \right] dy$$

3. Find the definite integral of $f(x) = \sin x$ from $x = 0°$ to $x = 90°$.

4. Find the definite integral of $f(x) = \sin x$ from $x = 0°$ to $x = 180°$.

5. Find the definite integral of $f(x) = \sin x$ from $x = 0°$ to $x = 270°$.

SOLUTIONS

1. Remember that the integral of a sum of functions is equal to the sum of the integrals of each of the individual functions. We also need to use both the rule for the indefinite integral of a variable raised to an integer power and the rule for the indefinite integral of a function multiplied by a constant. With these three principles in mind, proceed as follows:

$$\int (24y^5 + 15y^4 - 9y^2 + 4y - 4)\, dy$$

$$= \int 24y^5\, dy + \int 15y^4\, dy - \int 9y^2\, dy + \int 4y\, dy - \int 4\, dy$$

$$= 4y^6 + a_1 + 3y^5 + a_2 - 3y^3 + a_3 + 2y^2 + a_4 - 4y + a_5$$

The values a_1 through a_5 represent the constants of integration for each addend in this sum of functions. If we let them all add up to a single constant, a, then our final answer becomes:

$$4y^6 + 3y^5 - 3y^3 + 2y^2 - 4y + a$$

2. This looks formidable until we realize that it can be simplified to the indefinite integral of the solution to Problem 1. That is:

$$\int \left[\int (24y^5 + 15y^4 - 9y^2 + 4y - 4)\, dy \right] dy$$

$$= \int (4y^6 + 3y^5 - 3y^3 + 2y^2 - 4y + a)\, dy$$

$$= \int 4y^6\, dy + \int 3y^5\, dy - \int 3y^3\, dy + \int 2y^2\, dy - \int 4y\, dy + \int a\, dy$$

$$= (4/7)y^7 + b_1 + (1/2)y^6 + b_2 - (3/4)y^4 + b_3 + (2/3)y^3 + b_4 - 2y^2 + b_5 + ay + b_6$$

The values b_1 through b_6 represent the constants of integration for each addend in this sum of functions. If we let them all add up to a single constant, b, then our final answer becomes:

$$= (4/7)y^7 + (1/2)y^6 - (3/4)y^4 + (2/3)y^3 - 2y^2 + ay + b$$

Note that the constants a and b are not necessarily the same.

3. Before solving any integral that contains a trigonometric function, it's always a good idea to convert the independent variable to units of radians (rad). In this case, because $0° = 0$ rad and $90° = \pi/2$ rad, we are looking for the definite integral from $x = 0$ to $x = \pi/2$. First, find the indefinite integral Of the sine function from Table 15-1:

$$\int \sin x \, dx = -\cos x + c$$

We can consider the antiderivative $F(x)$ to be simply $-\cos x$, because when we find the definite integral, the constant c subtracts from itself and vanishes. The definite integral we seek can be expressed this way:

$$F(\pi/2) - F(0) = -\cos(\pi/2) - (-\cos 0)$$
$$= 0 - (-1)$$
$$= 0 + 1$$
$$= 1$$

4. Note that $0° = 0$ rad and $180° = \pi$ rad. We determined, in solving Problem 3, that the antiderivative $F(x)$ is equal to $-\cos x$. Therefore, the definite integral from 0 to π can be expressed as follows:

$$F(\pi) - F(0) = -\cos \pi - (-\cos 0)$$
$$= -(-1) - (-1)$$
$$= 1 + 1$$
$$= 2$$

5. Note that $0° = 0$ rad and $270° = 3\pi/2$ rad. We determined, in solving Problem 3, that the antiderivative $F(x)$ is equal to $-\cos x$. Therefore, the definite integral from 0 to $3\pi/2$ can be expressed as follows:

$$F(3\pi/2) - F(0) = -\cos(3\pi/2) - (-\cos 0)$$
$$= 0 - (-1)$$
$$= 0 + 1$$
$$= 1$$

Quiz

This is an "open book" quiz. You may refer to the text in this chapter. You may draw diagrams if that will help you visualize things. A good score is 8 correct. Answers are in the back of the book.

1. Figure 15-10 is a graph of the tangent function, $f(x) = \tan x$, over values of the domain ranging from -3π to 3π. Refer to Table 15-1 for the indefinite integral of the tangent function, and recall that the secant of a variable x is equal to the reciprocal of the cosine x:

$$\sec x = 1/(\cos x)$$

provided that x is not an odd multiple of $\pi/2$.

Based on this information, what happens to the area under the curve between $x = 0$ and $x = k$, as k increases from 0 to $\pi/2$?

(a) It grows positively, and reaches a certain positive finite value when $k = 1$.

(b) It grows negatively, and reaches a certain negative finite value when $k = 1$.

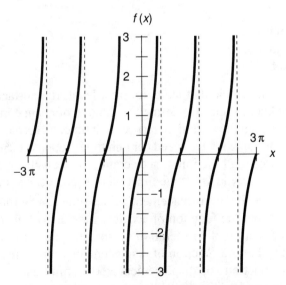

Fig. 15-10. Illustration for Quick Practice Problem and Solution 10.

(c) It grows positively without limit.

(d) It grows negatively without limit.

2. Based on Fig. 15-10 and Table 15-1, what is the definite integral of $f(x)$ with respect to x, from $x = 0$ to $x = \pi/2$? Express the answer to four decimal places. Consider $\pi = 3.14159$.

(a) −0.3466.

(b) 0.3466.

(c) 0.0000.

(d) It is not defined.

3. Based on Fig. 15-10 and Table 15-1, what is the definite integral of $f(x)$ with respect to x, from $x = -\pi/2$ to $x = \pi/2$? Express the answer to four decimal places. Consider $\pi = 3.14159$.

(a) −0.6931.

(b) 0.6931.

(c) 0.0000.

(d) It is not defined.

4. Based on Fig. 15-10 and Table 15-1, what is the definite integral of $f(x)$ with respect to x, from $x = 0$ to $x = \pi/4$? Express the answer to four decimal places. Consider $\pi = 3.14159$.

(a) 0.3466.

(b) 0.6931.

(c) −0.3466.

(d) −0.6931.

5. Consider a vehicle that starts up on a level, flat surface and moves in a straight line. Its speed is defined as the rate of change in the displacement (distance from the starting point) per unit time. Its acceleration is defined as the rate of change of the speed per unit time. Suppose the initial position, the initial speed, and the initial acceleration of the vehicle are all equal to 0. Also, suppose you know the function of speed versus time. Based on this,

(a) the function of displacement versus time can be found by integrating once, but the function of acceleration versus time cannot.

(b) the function of acceleration versus time can be found by integrating once, but the function of displacement versus time cannot.

(c) the functions of displacement versus time and acceleration versus time can both be found by integrating once.

(d) neither the function of displacement versus time nor the function of acceleration versus time can be found by integrating once.

6. Consider the scenario described in Question 5. Suppose the initial position, the initial speed, and the initial acceleration of the vehicle are all equal to 0. Also, suppose you know the function of acceleration versus time. Based on this,

    (a) the function of displacement versus time can be found by integrating one or more times, but the function of speed versus time cannot.
    (b) the function of speed versus time can be found by integrating one or more times, but the function of displacement versus time cannot.
    (c) the functions of displacement versus time and speed versus time can both be found by integrating one or more times.
    (d) neither the function of displacement versus time nor the function of speed versus time can be found by integrating one or more times.

7. Consider again the scenario described in Question 5. Suppose the initial position, the initial speed, and the initial acceleration of the vehicle are all equal to 0. Also, suppose you know the function of displacement versus time. Based on this,

    (a) the function of acceleration versus time can be found by integrating once, but the function of speed versus time cannot.
    (b) the function of speed versus time can be found by integrating once, but the function of acceleration versus time cannot.
    (c) the functions of acceleration versus time and speed versus time can both be found by integrating once.
    (d) neither the function of acceleration versus time nor the function of speed versus time can be found by integrating once.

8. What is the average value of $f(x) = 6x^2$ over the interval $0 \leq x < 2$?

    (a) 2.
    (b) 4.
    (c) 8.
    (d) 12.

9. What is the average value of $g(z) = \sin z$ over the interval $-\pi/2 < z \leq \pi/2$?

    (a) 0.
    (b) $2/\pi$.
    (c) $-2/\pi$.
    (d) It is not defined.

10. What is the indefinite integral, with respect to a variable x, of the sum of three or more different functions of x?

   (a) The arithmetic mean of the indefinite integrals of each of the individual functions with respect to x.

   (b) The geometric mean of the indefinite integrals of each of the individual functions with respect to x.

   (c) It is impossible to define the integral of a sum of more than two functions.

   (d) None of the above statements (a), (b), or (c) is correct.

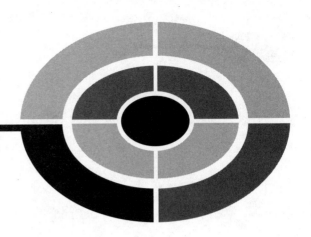

Final Exam

Do not refer to the text when taking this exam. You may draw diagrams or use a calculator if necessary. A good score is at least 75 answers (75% or more) correct. Answers are in the back of the book.

1. Suppose that the graphs of two equations in the Cartesian plane are identical; they coincide. This indicates that the pair of equations represented by the graphs

   (a) has no solution.
   (b) has exactly one solution.
   (c) has two distinct solutions.
   (d) has infinitely many solutions.
   (e) is nonlinear.

2. The equation $x^2 = -256$ has

   (a) two solutions, both real numbers.
   (b) two solutions, one a real number and the other a complex number.
   (c) two solutions, both of them complex numbers.
   (d) infinitely many solutions.
   (e) no solutions.

3. What is the value of $(-900)^{1/2}$?

   (a) 30.
   (b) −30.
   (c) $j30$.
   (d) ±30.
   (e) It is not defined.

4. Consider the following set of simultaneous linear equations in variables x, y, and z:

$$x + y + z - 3 = 0$$
$$-2x - 2y - 2z + 6 = 0$$
$$5x + 5y + 5z - 15 = 0$$

What can be said about the number of solutions here?

   (a) There are infinitely many solutions.
   (b) There are three distinct solutions.
   (c) There are two distinct solutions.
   (d) There is one unique solution.
   (e) There are no solutions.

5. Consider the following equation in two variables x and y:

$$4x + 5y - 7x - 21y = -4$$

This is an example of

   (a) a linear equation that is in standard form.
   (b) a linear equation that is not in standard form.
   (c) a quadratic equation that is in standard form.
   (d) a quadratic equation that is not in standard form.
   (e) a complex equation that is in standard form.

6. Consider the expression 7.898×10^7. Which of the following represents this quantity in plain decimal format?

   (a) 789,800,000
   (b) 78,980,000
   (c) 7,898,000
   (d) 0.0000007898
   (e) 0.00000007898

7. A pair of linear equations in x and y is inconsistent if any only if

   (a) x is a positive real number and y is a negative real number, or vice-versa.

   (b) x is an imaginary number and y is a real number, or vice-versa.

   (c) x and y are complex conjugates.

   (d) $x = 0$ and $y = 0$.

   (e) there exists no set $\{x, y\}$ that solves the pair of equations.

8. In Fig. Exam-1, the x axis

   (a) spans 3 orders of magnitude.

   (b) spans 7 orders of magnitude.

   (c) spans 10 orders of magnitude.

   (d) is linear.

   (e) represents an undefined range of quantities.

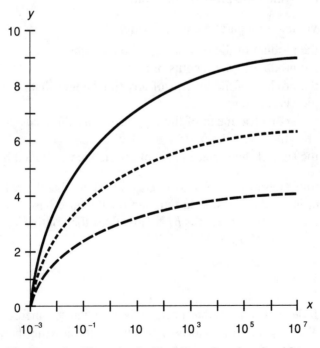

Fig. Exam-1. Illustration for Final Exam Questions 8 and 9.

9. In Fig. Exam-1, the y axis

   (a) spans 3 orders of magnitude.
   (b) spans 7 orders of magnitude.
   (c) spans 10 orders of magnitude.
   (d) is linear.
   (e) represents an undefined range of quantities.

10. Consider the following general equation, which holds for all real numbers a_1, a_2, a_3, b_1, b_2, and b_3, and where j is equal to the positive square root of -1:

$$[(a_1 + jb_1) + (a_2 + jb_2)] + (a_3 + jb_3)$$
$$= (a_1 + jb_1) + [(a_2 + jb_2) + (a_3 + jb_3)]$$

This is an expression of

   (a) the property of additive inverses.
   (b) the associative property of addition.
   (c) the commutative property of addition.
   (d) the transitive property of addition.
   (e) the symmetric property of addition.

11. The volume of a parallelepiped is equal to

   (a) the product of the base area and the height.
   (b) the product of the lengths of all the edges.
   (c) the product of the lengths of any three edges that converge at a single vertex point.
   (d) the geometric mean of the lengths of any three edges that converge at a single vertex point.
   (e) the sum of the surface areas of all the faces, divided by 6.

12. Imagine a tetrahedron X whose base is a triangle with area A, and whose height is h. Also consider a tetrahedron Y whose base is a triangle with area A, and whose height is $h/2$. How does the volume V_X of X compare with the volume V_Y of Y?

   (a) $V_X = 2^{1/2}V_Y$
   (b) $V_X = 2V_Y$
   (c) $V_X = 4V_Y$
   (d) It depends on whether or not X is a regular tetrahedron.
   (e) There is no way to answer this without knowing the actual dimensions of X.

13. Consider the following equation:

$$3x + 5 = 5x - 8$$

This is an example of

(a) a first-order equation in standard form.
(b) a first-order equation not in standard form.
(c) a quadratic equation in standard form.
(d) a quadratic equation not in standard form.
(e) a composite equation.

14. Consider the following equation:

$$6x - 5 = 8x + 3$$

What is the solution set for this equation?

(a) $\{4\}$
(b) $\{-4\}$
(c) $\{-4, 4\}$
(d) $\{-j4, j4\}$
(e) $\varnothing$

15. In Fig. Exam-2, suppose the right ascension of point P is 12 h. Then the angle labeled q represents

(a) the azimuth of point P.
(b) the celestial longitude of point P.
(c) the range of point P.
(d) the projection angle of point P.
(e) the declination of point P.

16. Consider an ellipsoid with radii of r_1, r_2, and r_3. If r_3 is cut in half while r_1 and r_2 do not change, what happens to the volume of the ellipse?

(a) It decreases by a factor of the square root of 2.
(b) It decreases by a factor of 2.
(c) It decreases by a factor of 4.
(d) It decreases by a factor of 8.
(e) We can't answer this without knowing the values of the radii to begin with.

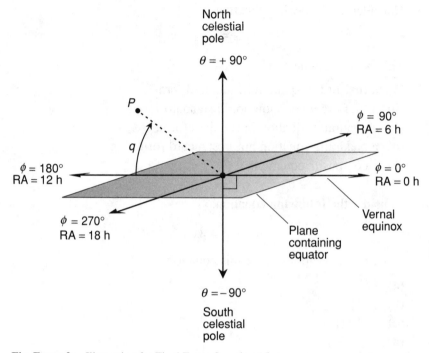

Fig. Exam-2. Illustration for Final Exam Question 15.

17. Consider the function $f(x) = 1/x$. As x increases positively without limit, what happens to the value of the first derivative of f?

    (a) It increases positively without limit.
    (b) It maintains negative values, but approaches 0.
    (c) It maintains positive values, but approaches 0.
    (d) It increases negatively without limit.
    (e) We cannot answer this question without more information.

18. The equation $x^2 + y^2 = -256$ has

    (a) two solutions, both real numbers.
    (b) two solutions, one a real number and the other a complex number.
    (c) two solutions, both of them complex numbers.
    (d) infinitely many solutions.
    (e) no solutions.

19. Consider the following expression in Boolean algebra:

$$A + -B \times C = D + E$$

Which of the following expressions is equivalent to this, and correctly clarifies the order in which the operations should be performed?

(a) $(A + (-(B \times C))) = (D + E)$
(b) $(A + (-(B \times C) = D) + E)$
(c) $(A + ((-B) \times C)) = (D + E)$
(d) $((A + (-B)) \times C) = (D + E)$
(e) $(A + (-B \times (C = D)) + E))$

20. What is the decimal-number value of octal 358?

(a) 358.
(b) 101100110.
(c) 1606.
(d) 96.
(e) It is not defined, because 358 is not a legitimate octal number.

21. The symbolic Boolean name for the logical operation that translates to "either/or" is

(a) IOR.
(b) EOR.
(c) NOR.
(d) XOR.
(e) OR.

22. Imagine a rectangle with an interior area of 100.000 m². We are not told the exact dimensions, but only the fact that it is a plane rectangle. What is the possible range of lengths, d, of a corner-to-corner diagonal of this rectangle?

(a) $d = 14.1421$ m.
(b) 10.0000 m $< d \le 14.1421$ m.
(c) $d \ge 14.1421$ m.
(d) $d > 10.0000$ m.
(e) We must have more information to answer this.

23. Many calculators lack cotangent function buttons, although they have buttons for the sine and cosine functions. How can you find the cotangent of an angle with such a calculator?

    (a) Divide 1 by the sine of the angle.
    (b) Divide 1 by the cosine of the angle.
    (c) Divide the sine of the angle by the cosine of the angle.
    (d) Divide the cosine of the angle by the sine of the angle.
    (e) You can't find the cotangent of an angle with such a calculator.

24. Examine Fig. Exam-3. The polygon is a trapezoid. The top and bottom (horizontal) sides are parallel with lengths as indicated, and are separated by the distance shown. What is the perimeter of this trapezoid, based on the information given in the diagram?

    (a) 22.8 m
    (b) 39.6 m
    (c) 51.2 m
    (d) 64.7 m
    (e) It cannot be determined from the information provided.

25. What is the interior area of the trapezoid shown in Fig. Exam-3, based on the information given in the diagram, and accurate to two significant figures?

    (a) 89 m²
    (b) 44 m²

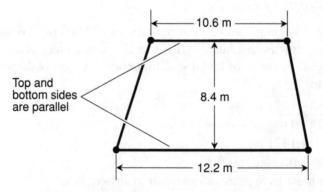

Fig. Exam-3.   Illustration for Final Exam Questions 24 and 25.

(c) 96 m^2

(d) 48 m^2

(e) It cannot be determined from the information provided.

26. In Cartesian 3-space, how far is the point (10,10,10) from the origin? Consider the coordinates exact, and express the answer to four significant figures.

    (a) 5.477 units.

    (b) 6.694 units.

    (c) 10.00 units.

    (d) 17.32 units.

    (e) 31.62 units.

27. The dot product of two vectors is always

    (a) a natural number.

    (b) an integer.

    (c) a rational number.

    (d) a real number.

    (e) None of the above

28. The common logarithms of −100 and 100 differ by

    (a) 0.

    (b) 1.

    (c) 2.

    (d) 4.

    (e) No amount, because the common logarithm of −100 is not defined.

29. Consider the following function f of a variable x:

$$f(x) = -3x^4 + 5x^3 + 4$$

What is $f''(x)$?

    (a) $f''(x) = -12x^3 + 15x^2$

    (b) $f''(x) = 12x^3 - 15x^2$

    (c) $f''(x) = 36x^2 - 30x$

    (d) $f''(x) = -36x^2 + 30x$

    (e) $f''(x)$ cannot be defined unless we know the constant of integration.

30. Suppose point P is an inflection point on a curve that represents a continuous function $y = f(x)$. Which, if any, of the following statements (a), (b), (c), or (d) is true in general?

    (a) The sense of concavity of the curve reverses at point P.
    (b) Point P is a local minimum of f.
    (c) Point P is a local maximum of f.
    (d) The derivative of f is equal to 0 at point P.
    (e) None of the above statements (a), (b), (c), and (d) are true in general.

31. How many definite integrals can the following function have?

    $$f(x) = 3x^3 + 4x^2 + 3x + 2$$

    (a) One.
    (b) Two.
    (c) Three.
    (d) Four.
    (e) Infinitely many.

32. How many definite integrals does the following function have for values of x between, and including, 0 and 1?

    $$f(x) = 3x^3 + 4x^2 + 3x + 2$$

    (a) One.
    (b) Two.
    (c) Three.
    (d) Four.
    (e) Infinitely many, because the constant of integration can be any real number between 0 and 1.

33. Consider the function $f(x) = 1/x$. As x remains positive but approaches 0, what happens to the value of the first derivative of f?

    (a) It increases positively without limit.
    (b) It increases negatively without limit.
    (e) It maintains positive values, but approaches 0.
    (d) It maintains negative values, but approaches 0.
    (e) We cannot answer this question without more information.

34. Imagine a function $h(x)$ that has a graph in the xy-plane that you don't recognize as any familiar class of curve. Suppose we are told that the area between the curve and the x axis, for the span of values of x between $x = -5$ and $x = 5$, is equal to -100. What is the average value of the function $h(x)$ over the interval from $x = -5$ to $x = 5$?

    (a) We are not given the values of $h(-5)$ and $h(5)$. Unless we know these values specifically, there is no way for us to determine the average value of $h(x)$ over this interval.

    (b) We are not given the values of $H(-5)$ and $H(5)$, where H is the anti-derivative of h. Unless we know these values specifically, there is no way for us to determine the average value of $h(x)$ over this interval.

    (c) The area is given as negative, but there is no such thing as negative area. Thus, the average value of the function over this interval is not defined.

    (d) The average value of the function $h(x)$ over the interval from $x = -5$ to $x = 5$ is equal to -10.

    (e) The average value of the function $h(x)$ over the interval from $x = -5$ to $x = 5$ is equal to 0.

35. Consider a vector $\mathbf{a} = 0\mathbf{i} + 6\mathbf{j} + 8\mathbf{k}$ in Cartesian xyz-space. If the originating point of this vector is the coordinate origin $(0,0,0)$, then $|\mathbf{a}|$ is equal to

    (a) 0.
    (b) 6.
    (c) 8.
    (d) 10.
    (e) a value that requires more information to be calculated.

36. Imagine a vertical utility pole 17.4 m tall that stands in a flat, level field. The shadow of the pole, cast by the sun, is 44.5 m long. What is the area, in square meters (m^2), of the triangle defined by the base of the pole, the top of the pole, and the end of the pole's shadow?

    (a) $31.0 \ m^2$
    (b) $61.9 \ m^2$
    (c) $387 \ m^2$
    (d) $774 \ m^2$
    (e) $1549 \ m^2$

37. Imagine a vertical utility pole 17.4 m tall that stands in a flat, level field. The shadow of the pole, cast by the sun, is 44.5 m long. What is the distance, in meters (m), from the top of the pole to the tip of its shadow?

    (a) 31.0 m
    (b) 47.8 m
    (c) 61.9 m
    (d) 95.6 m
    (e) More information is necessary to determine this.

38. In Newtonian physics, the displacement function is

    (a) the derivative of the acceleration function with respect to time.
    (b) the derivative of the speed function with respect to time.
    (c) the indefinite integral of the acceleration function with respect to time.
    (d) the indefinite integral of the speed function with respect to time.
    (e) None of the above

39. Consider the following expression in Boolean algebra:

    $$(X \times Y) + (X \times Z) = X \times (Y + Z)$$

    This statement is a general expression of

    (a) the associative law of conjunction with respect to disjunction.
    (b) the commutative law of conjunction with respect to disjunction.
    (c) the commutative law of disjunction with respect to conjunction.
    (d) the distributive law of conjunction with respect to disjunction.
    (e) a logically invalid statement, because it is not in general true.

40. Refer to Fig. Exam-4. What is the sum of the complex numbers represented by the points P and Q?

    (a) $3 + j0$.
    (b) $-3 + j0$.
    (c) $7 + j6$.
    (d) $7 - j6$.
    (e) $-7 - j6$.

41. Refer to Fig. Exam-4. What is the product of the complex numbers represented by the points P and Q?

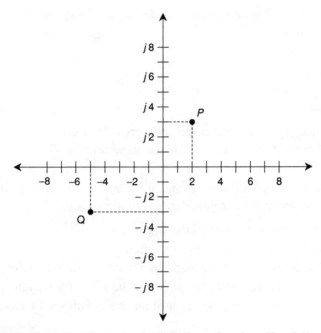

Fig. Exam-4. Illustration for Final Exam Questions 40 and 41.

    (a) $-1 - j21$.
    (b) $-1 + j0$.
    (c) $-10 + j9$.
    (d) $10 - j9$.
    (e) It is not defined.

42. Consider the product $4.5101 \times 256 \times 82.22 \times 7.6 \times e$, where e is the natural logarithm base, a well-known irrational-number constant. How many significant figures can we claim after we have multiplied these numbers?

    (a) As many as we want.
    (b) 5.
    (c) 4.
    (d) 3.
    (e) 2.

43. How many radians equal one angular degree? Express the answer to five significant figures.

    (a) 0.31831.
    (b) 0.15915.
    (c) 0.017453.
    (d) 0.0087266.
    (e) This question cannot be answered, because radians and angular degrees do not represent the same parameter.

44. How can you find the cosecant of an angle with a calculator that has only sine and cosine trigonometric function buttons?

    (a) Divide 1 by the sine of the angle.
    (b) Divide 1 by the cosine of the angle.
    (c) Divide the sine of the angle by the cosine of the angle.
    (d) Divide the cosine of the angle by the sine of the angle.
    (e) You can't find the cosecant of an angle with such a calculator.

45. Imagine a triangle with a perimeter of 30.0 m. We are not told its specific dimensions, but only the fact that it is a plane triangle. What is the possible range of interior areas, A, of this triangle?

    (a) $0 < A \leq 21.7$ m^2.
    (b) $0 < A \leq 43.3$ m^2.
    (c) $0 < A \leq 90.0$ m^2.
    (d) $0 < A \leq 120$ m^2.
    (e) $0 < A \leq 150$ m^2.

46. Which of the following is *not* a rational number?

    (a) 23.23
    (b) $(-23)/(-99)$
    (c) The natural logarithm of 1
    (d) The positive square root of 16
    (e) The cube root of 7

47. In the form of cylindrical coordinates used by aviators and navigators, the term *azimuth*, in reference to a target point, refers to

(a) the actual radius, or distance from the origin to the target.

(b) the compass direction, in degrees, expressed clockwise from north, for the projection of the target onto the horizontal plane.

(c) the compass direction, in radians, measured counterclockwise from north, for the projection of the target onto the horizontal plane.

(d) the distance from the origin to the projection of the target onto the horizontal plane.

(e) the angle between the horizontal plane and a ray connecting the origin with the target.

48. Consider a continuous function f in the xy-plane, such that $y = f(x)$. Suppose P is a point on the curve representing the function f. Further suppose that $f'(x) = 0$ and $f''(x) = -3$ at point P. From this information, we can be certain that

(a) point P is a local minimum of f.

(b) point P is a local maximum of f.

(c) point P is an inflection point of f.

(d) All of the above statements (a), (b), and (c) are true.

(e) None of the above statements (a), (b), and (c) are true.

49. Consider the three standard unit vectors $\mathbf{i}$, $\mathbf{j}$, and $\mathbf{k}$ in Cartesian xyz-space. Which of the following statements (a), (b), (c), or (d), if any, is false?

(a) Each of these vectors has a magnitude equal to 1.

(b) Each of these vectors is perpendicular to both of the other two.

(c) The cross product $\mathbf{i} \times \mathbf{j}$ is equal to the vector $\mathbf{k}$.

(d) The magnitude of $\mathbf{i} + \mathbf{j} + \mathbf{k}$ is equal to the square root of 3.

(e) All of the above statements (a), (b), (c), and (d) are true.

50. Imagine an extremely thin, flat, circular disk. If its radius is reduced by a factor of 16, then its surface area

(a) is reduced by a factor of 4096.

(b) is reduced by a factor of 256.

(c) is reduced by a factor of 16.

(d) is reduced by a factor of 4.

(e) is reduced by a factor of 2.

51. Suppose you have a pole that is 4.000 m high, and you also have some canvas. You want to build a temporary shelter on a level field, in the shape of a right circular cone. You want the radius of the cone to be 3.000 m. Recall the formula for the lateral surface area of a right circular cone. If r is the radius of the base and h is the height of the cone at the center, then the lateral surface area S_L is given by:

$$S_L = \pi r (r^2 + h^2)^{1/2}$$

How much canvas will you need to build your shelter, rounded up to the next higher square meter? Consider $\pi = 3.14159$.

(a) This cannot be determined unless we are told the slant height of the cone.
(b) 16 m^2.
(c) 47 m^2.
(d) 48 m^2.
(e) 95 m^2.

52. What is the absolute value of $-33 - j44$?

(a) $33 + j44$.
(b) 33.
(c) 44.
(d) 55.
(e) 77.

53. Suppose that the graphs of two equations in the Cartesian plane are parallel, straight lines. This indicates that the pair of equations represented by the graphs

(a) has no common solution.
(b) has exactly one common solution.
(c) has two distinct common solutions.
(d) has infinitely many common solutions.
(e) is nonlinear.

54. In Fig. Exam-5, the quantity s/t represents

(a) $\sin \theta$.
(b) $\cos \theta$.
(c) $\tan \theta$.
(d) $\cot \theta$.
(e) $\sec \theta$.

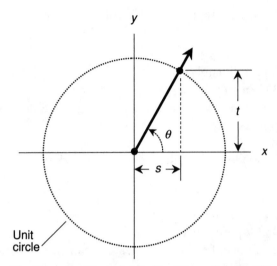

Fig. Exam-5. Illustration for Final Exam Questions 54 and 55.

55. In Fig. Exam-5, the quantity $1/s$ represents

    (a) $\sin \theta$.

    (b) $\cos \theta$.

    (c) $\tan \theta$.

    (d) $\cot \theta$.

    (e) $\sec \theta$.

56. Consider an ellipsoid with radii of r_1, r_2, and r_3. If r_1 and r_3 are both cut in half while r_2 does not change, what happens to the volume of the ellipse?

    (a) It decreases by a factor of the square root of 2.

    (b) It decreases by a factor of 2.

    (c) It decreases by a factor of 4.

    (d) It decreases by a factor of 8.

    (e) We can't answer this without knowing the values of the radii to begin with.

57. What is the arithmetic mean of 10, 100, and 1000?

   (a) 100.
   (b) 370.
   (c) 667 (approximately).
   (d) 333,333 (approximately).
   (e) There is none, because arithmetic means are defined only for pairs of numbers.

58. What is the geometric mean of 10, 100, and 1000?

   (a) 100.
   (b) 370.
   (c) 667 (approximately).
   (d) 333,333 (approximately).
   (e) There is none, because geometric means are defined only for pairs of numbers.

59. Which of the following equations (a), (b), (c), or (d), if any, is false?

   (a) $1/(10^x) = 10^{-x}$
   (b) $5^{(x/y)} = (5^x)^{(1/y)}$
   (c) $(3^x)^y = 3^{(xy)}$
   (d) $e^x e^y = e^{(x+y)}$
   (e) All of the above statements are true.

60. Imagine a pair of equations in two variables. Suppose that their graphs appear as shown in Fig. Exam-6, and that the graphs extend forever in the directions implied by the arrows. From this, it is apparent that the pair of equations

   (a) has no common solution.
   (b) has exactly one common solution.
   (c) has two distinct common solutions.
   (d) has more than two distinct common solutions, but not infinitely many.
   (e) has infinitely many distinct common solutions.

61. Consider two numbers p and q, such that $p = 10,000q$. By how many orders of magnitude do p and q differ?

   (a) 5.
   (b) 4.
   (c) 3.
   (d) 2.
   (e) More information is necessary to answer this.

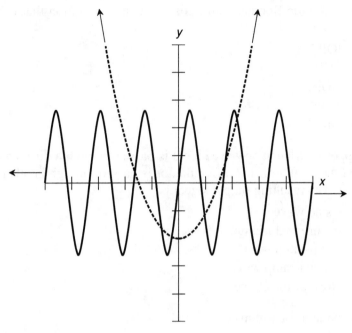

Fig. Exam-6. Illustration for Final Exam Question 60.

62. Consider two numbers r and s, such that $r = s/10,000$. By how many orders of magnitude do r and s differ?

    (a) 5.
    (b) 4.
    (c) 3.
    (d) 2.
    (e) More information is necessary to answer this.

63. Imagine a utility pole 35 m high, standing upright (perfectly vertical) in a flat, horizontal field. Suppose the shadow of the pole, cast by the sun, is 70 m long. What is the angle of the sun with respect to the zenith (the point in the sky directly overhead), to the nearest degree?

    (a) 30°.
    (b) 60°.
    (c) 27°.
    (d) 63°.
    (e) This angle cannot be calculated without more information.

64. The symbolic Boolean name for the logical operation that translates to "inclusive or" is

    (a) IOR.
    (b) EOR.
    (c) NOR.
    (d) XOR.
    (e) OR.

65. Suppose a triangular wave signal is passed through a differentiator, and then the output of this differentiator is passed through another differentiator. The resulting output signal is

    (a) a sine wave.
    (b) the original triangular wave.
    (c) an up-ramp wave.
    (d) a down-ramp wave.
    (e) None of the above

66. The natural logarithms of e^{-4} and e^4 differ by

    (a) 2.
    (b) 4.
    (c) 8.
    (d) 16.
    (e) No amount, because the natural logarithm of e^{-4} is not defined.

67. The cross product of a vector with itself is

    (a) the vector $\mathbf{i}^2 + \mathbf{j}^2 + \mathbf{k}^2$.
    (b) the vector $\mathbf{i} + \mathbf{j} + \mathbf{k}$.
    (c) the zero vector.
    (d) the scalar quantity 1.
    (e) the scalar quantity 0.

68. Examine Fig. Exam-7. What is the equation represented by the upper line in this graph?

    (a) $y = (3/4)x - 4$
    (b) $y = (3/4)x - 3$
    (c) $y = (3/4)x + 4$
    (d) $y = (3/4)x + 3$
    (e) More information is needed to answer this.

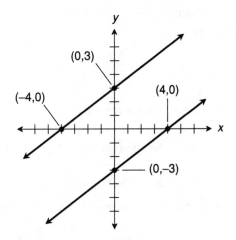

Fig. Exam-7. Illustration for Final Exam
Questions 68 through 70.

69. What is the equation represented by the lower line in Fig. Exam-7?

    (a) $y = (3/4)x - 4$
    (b) $y = (3/4)x - 3$
    (c) $y = (3/4)x + 4$
    (d) $y = (3/4)x + 3$
    (e) More information is needed to answer this.

70. How many common solutions are there to the pair of equations represented by the lines in Fig. Exam-7?

    (a) Infinitely many.
    (b) Two.
    (c) One.
    (d) None.
    (e) More information is needed to answer this.

71. The cardinality of the set $\{2, 4, 6, 8, 10, \ldots\}$ is

    (a) a positive integer.
    (b) aleph 0.
    (c) a transcendental number.
    (d) an imaginary number.
    (e) an irrational number.

72. Which of the following equations (a), (b), (c), or (d), if any, is false?
    (a) $\ln xy = \ln x + \ln y$
    (b) $\ln x = e^{-x}$
    (c) $\ln x^y = y \ln x$
    (d) $\ln (1/x) = -\ln x$
    (e) All of the above statements are true.

73. Let x be a variable. Let c be the constant of integration. Consider this:

    $$\int x^n \, dx = [x^{(n+1)}/(n+1)] + c$$

    This formula holds true for all integer values of n except
    (a) $n = 0$.
    (b) $n = 1$.
    (c) $n = -1$.
    (d) $n = -1$ and $n = 1$.
    (e) Forget the exception! The formula holds true for all integers.

74. Imagine a deflating, spherical balloon. If its radius is reduced by a factor of 16, then its surface area
    (a) is reduced by a factor of 4096.
    (b) is reduced by a factor of 256.
    (c) is reduced by a factor of 16.
    (d) is reduced by a factor of 4.
    (e) is reduced by a factor of 2.

75. In a single-variable quintic equation, what is the largest value of the exponent to which the variable is raised?
    (a) 2.
    (b) 3.
    (c) 4.
    (d) 5.
    (e) It depends on the number of solutions the equation has.

76. Consider the equation $x^y = z$ (that is, x raised to the yth power equals z), where x and z are positive real numbers, and y can be any real number. In this situation,

(a) The base-x logarithm of z is equal to y.
(b) The base-y logarithm of z is equal to x.
(c) The base-z logarithm of y is equal to x.
(d) The base-z logarithm of x is equal to y.
(e) None of the above statements is true.

77. The value of $(-4)!$ is
   (a) 24.
   (b) −24.
   (c) 0.
   (d) $j24$.
   (e) undefined.

78. The cross product of two vectors is always
   (a) a natural number.
   (b) an integer.
   (c) a rational number.
   (d) a real number.
   (e) None of the above

79. Of what order is the following equation?

$$(x^2 + 6)(5x^2 - 5)(8x - 3) = 0$$

   (a) There is no way to answer this without more information.
   (b) Fifth order (quintic).
   (c) Fourth order (quartic).
   (d) Third order (cubic).
   (e) Second order (quadratic).

80. How can you find the secant of an angle with a calculator that has only sine and cosine trigonometric function buttons?
   (a) Divide 1 by the sine of the angle.
   (b) Divide 1 by the cosine of the angle.
   (c) Divide the sine of the angle by the cosine of the angle.
   (d) Divide the cosine of the angle by the sine of the angle.
   (e) You can't find the secant of an angle with such a calculator.

81. Which of the following represents the quantity −0.000002052 in plain-text scientific notation?

   (a) 2.052E−06
   (b) −2.052E−06
   (c) 2.052E+06
   (d) −2.052E+06
   (e) This quantity cannot be represented in plain-text scientific notation.

82. The *power gain* of an electronic circuit, in units called decibels (dB), is calculated according to the following formula:

$$\text{Gain (dB)} = 10 \log (P_{out}/P_{in})$$

where P_{out} is the output signal power and P_{in} is the input signal power, both specified in watts. Suppose the audio input to a circuit is 50 watts, and the output is 0.50 watts (the output is weaker than the input). What is the power gain of this circuit in decibels?

   (a) 14 dB.
   (b) 20 dB.
   (c) −14 dB.
   (d) −20 dB.
   (e) 0 dB.

83. In the situation shown by Fig. Exam-8, suppose each angular division represents 10°. In what direction does $\mathbf{a} \times \mathbf{b}$ point?

   (a) Straight up.
   (b) Straight down.
   (c) Generally eastward, horizontally.
   (d) In order to answer this, we must know the actual magnitudes of $\mathbf{a}$ and $\mathbf{b}$.
   (e) In no direction, because $\mathbf{a} \times \mathbf{b}$ is not a vector.

84. In the situation shown by Fig. Exam 8, let r_a be the radius of vector $\mathbf{a}$, and let r_b be the radius of vector $\mathbf{b}$. Suppose each angular division represents 10°. Then $\mathbf{a} \bullet \mathbf{b}$ is

   (a) equal to $r_a r_b$.
   (b) equal to $-r_a r_b$.
   (c) equal to 0.
   (d) equal to $r_a + r_b$.
   (e) equal to $(r_a^2 + r_b^2)^{1/2}$.

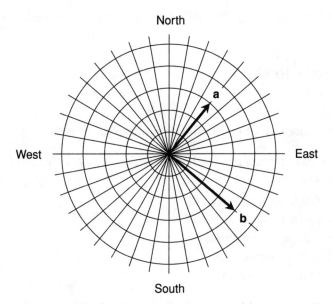

North

West East

South

Fig. Exam-8. Illustration for Final Exam Questions 83 and 84.

85. Suppose x is a positive real number. As x becomes smaller and smaller positively, approaching (but never quite reaching) 0, what happens to the value of $\ln x$?

    (a) It becomes larger and larger positively, without limit.
    (b) It approaches 1.
    (c) It approaches 0.
    (d) It becomes larger and larger negatively, without limit.
    (e) It oscillates endlessly between -1 and 1.

86. What is the product of 2.50×10^5 and 4.00×10^{-6} expressed to 3 significant figures?

    (a) 2.50×10^5
    (b) 4.00×10^{-6}
    (c) 1.00
    (d) 10.0×10^{-30}
    (e) It cannot be determined without more information.

87. What is the sum of 2.50×10^5 and 4.00×10^{-6} expressed to 3 significant figures?
    (a) 2.50×10^5
    (b) 4.00×10^{-6}
    (c) 1.00
    (d) 10.0×10^{-30}
    (e) It cannot be determined without more information.

88. In Boolean algebra, an atomic proposition always consists of either a logical constant or
    (a) a single logical variable.
    (b) a set of logical variables.
    (c) a logical conjunction.
    (d) a logical implication.
    (e) a logical equivalence.

89. The dot product of two vectors that are perpendicular to each other is
    (a) the vector $\mathbf{i}^2 + \mathbf{j}^2 + \mathbf{k}^2$.
    (b) the vector $\mathbf{i} + \mathbf{j} + \mathbf{k}$.
    (c) the zero vector.
    (d) the scalar quantity 1.
    (e) the scalar quantity 0.

90. Suppose an alternating current (AC) wave completes a cycle exactly once every 10 microseconds (µs). What is the frequency of this wave in kilohertz (kHz)?
    (a) 1 kHz.
    (b) 10 kHz.
    (c) 100 kHz.
    (d) 1000 kHz.
    (e) 10,000 kHz.

91. What general statement can be made about the number of solutions to a pair of linear equations in two variables?
    (a) There are no solutions.
    (b) There is one unique solution.
    (c) There are two distinct solutions.
    (d) There are infinitely many solutions.
    (e) We must see the equations before we can say how many solutions exist.

92. Suppose a radar set shows a target that is 100 km directly southeast of your location. What are the bearing and range of this target, expressed in navigator's polar coordinates? Express the bearing to the nearest degree, and the range to three significant figures.

    (a) The bearing is 315°, and the range is 100 km.
    (b) The bearing is 135°, and the range is 100 km.
    (c) The bearing is 225°, and the range is 70.7 km.
    (d) The bearing is 45°, and the range is 70.7 km.
    (e) More information is necessary to answer this.

93. Imagine a rectangle with an interior area of 100.000 m². We are not told the exact dimensions, but only the fact that it is a plane rectangle. What is the possible range of perimeters, B, of this rectangle?

    (a) $B = 40.0000$ m.
    (b) 20.0000 m $< B \leq 40.0000$ m.
    (c) $B \leq 40.0000$ m.
    (d) $B \geq 40.0000$ m.
    (e) We must have more information to answer this.

94. Suppose you stand in a flat field and fly a kite. The wind blows directly from the east. The point on the ground directly below the kite is 600 m away from you, and the kite is 800 m above the ground. If your body represents the origin of a spherical coordinate system, what is the radius coordinate of the kite in kilometers (km)?

    (a) 0.693 km
    (b) 0.700 km
    (c) 1.00 km
    (d) 1.40 km
    (e) There is no way to tell without more information.

95. Consider the following quadratic equation:

$$2.3x^2 + 3.3x - 10.5 = 0$$

What can be said about the solutions to this equation without actually solving it?

    (a) There is a single real-number solution.
    (b) There are two distinct real-number solutions.
    (c) There is a single complex-number solution.
    (d) There are two distinct complex-number solutions.
    (e) There are no solutions.

96. Consider two sets: $S = \{2, 4, 6, 8, \ldots\}$ and $T = \{1, 3, 5, 7, \ldots\}$. What is $S \cup T$?

    (a) The set of all positive integers.
    (b) The set of all rational numbers.
    (c) The set of all real numbers.
    (d) The set of all complex numbers.
    (e) The null set.

97. Which of the graphs in Fig. Exam-9 is or are such that y is a function of x within the portions of the domain and range as shown?

    (a) Graph A only.
    (b) Graph B only.
    (c) Graph C only.
    (d) Graphs A and B.
    (e) Graphs A and C.

98. Which of the graphs in Fig. Exam-9 is or are such that x is a function of y within the portions of the domain and range as shown?

    (a) Graph A only.
    (b) Graph B only.
    (c) Graph C only.
    (d) Graphs A and B.
    (e) Graphs A and C.

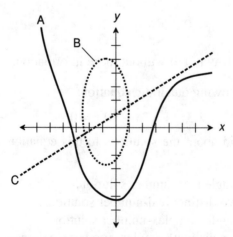

Fig. Exam-9. Illustration for Final Exam Questions 97 and 98.

99. In Newtonian physics, the acceleration function is
    (a) the derivative of the displacement function with respect to time.
    (b) the derivative of the speed function with respect to time.
    (c) the indefinite integral of the displacement function with respect to time.
    (d) the indefinite integral of the speed function with respect to time.
    (e) None of the above

100. The surface area, A, of a rectangular prism having edges of lengths r, s, and t is given by which of the following formulas?
    (a) $A = r + s + t$
    (b) $A = rst$
    (c) $A = 2rs + 2rt + 2st$
    (d) $A = r^2 + s^2 + t^2$
    (e) $A = r^2 s^2 t^2$

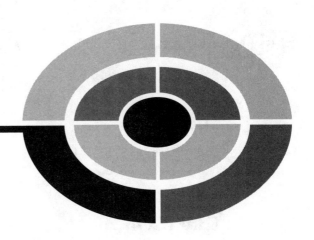

Answers to Quiz and Exam Questions

CHAPTER 1

1. b	2. b	3. b	4. d	5. a
6. b	7. c	8. d	9. a	10. c

CHAPTER 2

1. c	2. c	3. a	4. d	5. a
6. b	7. d	8. c	9. b	10. c

CHAPTER 3

1. d	2. a	3. b	4. c	5. c
6. c	7. c	8. c	9. d	10. c

CHAPTER 4

1. d	2. b	3. b	4. c	5. d
6. d	7. a	8. d	9. a	10. c

CHAPTER 5

1. c	2. a	3. d	4. a	5. c
6. b	7. d	8. b	9. c	10. d

CHAPTER 6

1. c	2. b	3. a	4. b	5. c
6. b	7. b	8. d	9. a	10. d

CHAPTER 7

1. a	2. b	3. b	4. c	5. a
6. b	7. d	8. a	9. b	10. d

CHAPTER 8

1. b	2. c	3. a	4. d	5. d
6. a	7. a	8. d	9. c	10. a

CHAPTER 9

1. a	2. a	3. d	4. d	5. d
6. d	7. b	8. b	9. a	10. c

CHAPTER 10

1. b   2. a   3. d   4. c   5. a
6. a   7. b   8. b   9. d   10. d

CHAPTER 11

1. d   2. a   3. c   4. b   5. d
6. b   7. a   8. d   9. c   10. d

CHAPTER 12

1. d   2. a   3. b   4. a   5. a
6. c   7. b   8. b   9. c   10. b

CHAPTER 13

1. d   2. c   3. d   4. d   5. b
6. c   7. b   8. d   9. d   10. c

CHAPTER 14

1. c   2. b   3. c   4. a   5. d
6. c   7. d   8. d   9. d   10. c

CHAPTER 15

1. c   2. d   3. d   4. a   5. a
6. c   7. d   8. c   9. a   10. d

FINAL EXAM

1. d	2. c	3. c	4. a	5. b
6. b	7. e	8. c	9. d	10. b
11. a	12. b	13. b	14. b	15. e
16. b	17. b	18. d	19. c	20. e
21. d	22. c	23. d	24. e	25. c
26. d	27. d	28. e	29. d	30. a
31. e	32. a	33. b	34. d	35. d
36. c	37. b	38. d	39. d	40. b
41. a	42. e	43. c	44. a	45. b
46. e	47. b	48. b	49. e	50. b
51. d	52. d	53. a	54. d	55. e
56. c	57. b	58. a	59. e	60. d
61. b	62. b	63. d	64. e	65. e
66. c	67. c	68. d	69. b	70. d
71. b	72. b	73. c	74. b	75. d
76. a	77. e	78. e	79. b	80. b
81. b	82. d	83. b	84. c	85. d
86. c	87. a	88. a	89. e	90. c
91. e	92. b	93. d	94. c	95. b
96. a	97. e	98. c	99. b	100. c

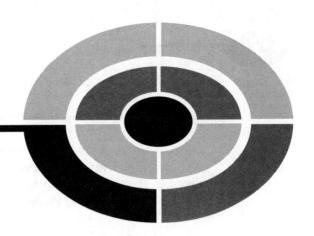

Suggested Additional Reading

Bluman, A. *Math Word Problems Demystified*. New York: McGraw-Hill, 2005.

Bluman, A. *Pre-Algebra Demystified*. New York: McGraw-Hill, 2004.

Crowhurst, N. and Gibilisco, S. *Mastering Technical Mathematics, 2nd edition*. New York: McGraw-Hill, 1999.

Gibilisco, S. *Everyday Math Demystified*. New York: McGraw-Hill, 2004.

Gibilisco, S. *Geometry Demystified*. New York: McGraw-Hill, 2003.

Gibilisco, S. *Statistics Demystified*. New York: McGraw-Hill, 2004.

Gibilisco, S. *Trigonometry Demystified*. New York: McGraw-Hill, 2003.

Huettenmueller, R. *Algebra Demystified*. New York: McGraw-Hill, 2003.

Huettenmueller, R. *College Algebra Demystified*. New York: McGraw-Hill, 2004.

Huettenmueller, R. *Precalculus Demystified*. New York: McGraw-Hill, 2005.

Krantz, S. *Calculus Demystified*. New York: McGraw-Hill, 2003.

Krantz, S. *Differential Equations Demystified*. New York: McGraw-Hill, 2004.

Olive, J. *Maths: A Student's Survival Guide, 2nd edition*. Cambridge, England: Cambridge University Press, 2003.

Shankar, R. *Basic Training in Mathematics: A Fitness Program for Science Students*. New York: Plenum Publishing Corporation, 1995.

INDEX

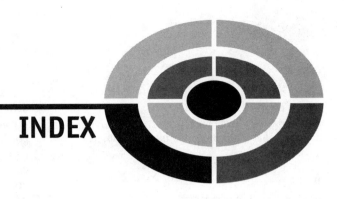

A

abscissa, 73, 231
absolute value of complex number, 18, 24
acceleration, 354–358
acute angle, 228
addition
   and significant figures, 64–65
   associative law of, 31
   commutative law of, 31–32
   of vectors, 272–274
   using scientific notation, 59–60, 64–65
addition method
   for solving 2-by-2 linear equations, 145–148
additive
   identity element, 30
   inverse, 30, 287
adjacent side, 236
aleph, 16–17
altitude, in cylindrical coordinates, 109
amplitude of wave function, 323–324, 349
angle
   acute, 228
   double, 242
   negative, 240–241
   nonstandard, 233–234
   notation for, 235

   obtuse, 228
   of intersection between curves, 317–318
   right, 164–165
angular
   degree, 228–229
   difference, 242
   radian, 228–229
   sum, 242
antecedent, 211
antiderivative, definition of, 338
antilogarithm, 292
apex angle of circular sector, 177–178
approximate-equality symbol, 62
approximation, 42–43
arc
   degree of, 228–229
   minute of, 95, 228–229
   second of, 95, 228–229
arithmetic mean, 35
associative law
   of addition, 31
   of conjunction, 217–218
   of disjunction, 217–218
   of multiplication, 32
   of vector addition, 272–274
   of vector-scalar multiplication, 274

astronomical unit (AU), 117
atomic proposition, 212
AU. *See* astronomical unit
average value of function over interval, 356–357
azimuth, 87, 112

B

back-end point of vector, 254
base-2 number system, 11–13
base-8 number system, 11–13
base-10
    exponential, 290–291
    logarithm, 284–285
    number system, 10, 12–13
base-16 number system, 12–13
base-*e*
    exponential, 290–293
    logarithm, 285–286
bearing, 87
binary number system, 11–13
Boolean algebra, 207–228

C

calculus
    differential, 305–335
    Fundamental Theorem of, 339–340
    integral, 337–364
Cantor, Georg, 17
cardinal number, transfinite, 16–17
cardinality of set, 5
Cartesian 3-space
    axes in, 107
    definition of, 105–106
    distance between points in, 107–108
    orientation of axes in, 260
    origin in, 107
    variables in, 107
    vectors in, 259–264
Cartesian 4-space, 114–115
Cartesian n-space
    definition of, 118
    distance between points in, 119
Cartesian plane
    abscissa in, 73

coordinate conversions to and from, 89–91
    definition of, 71–72
    distance between points in, 73–74
    graphs in, 74–80
    ordinate in, 73
    origin in, 73
    vectors in, 251–256
Cartesian time-space, 115–118
celestial coordinates, 94–95, 98–99, 112–113
chain rule for derivatives, 313
circle
    circumference of, 173–174
    interior area of, 173–175
circular function
    primary, 229–232
    secondary, 232–234
circular sector
    apex angle of, 177–178
    interior area of, 178
    perimeter of, 177–178
circumference of circle, 173–174
coefficient
    as multiple of variable, 244
    in scientific notation, 52
coincident sets, 3
common exponential
    definition of, 290
common logarithm
    definition of, 284–285
    in terms of natural logarithm, 288
commutative law
    negative, of cross product, 272–273
    of addition, 31–32
    of conjunction, 217
    of disjunction, 217
    of dot product, 272
    of multiplication, 31
    of vector addition, 272
    of vector-scalar multiplication, 272
complex number
    absolute value of, 18, 24
    conjugates, 22
    definition of, 18
    equality of, 21

operations with, 21
plane, 19
properties of, 19–24
component vector, 270–271
compound statement, 208
concave downward, 320
concave upward, 320
concavity, sense of, 320
cone
    definition of, 191–192
    frustum, surface area of, 193–194
    frustum, volume of, 194
    right circular, definition of, 192
    right circular, surface area of, 193
    right circular, volume of, 193
    slant circular, volume of, 194–195
conjunction, logical, 208–209, 213, 219
consequent, 211
constant
    function, 311
    of integration, 338–339
continuous function, 308
Continuum Hypothesis, 16, 19
contradiction, 217
contrapositive, 218–219
coordinates
    Cartesian 3-space, 105–108
    Cartesian 4-space, 114–115
    Cartesian plane, 73–80, 89–91
    celestial, 94–95
    conversions, among different systems, 89–92
    cylindrical, 109–110, 113
    declination and right ascension, 94–95, 99
    hyperspace, 113–119
    latitude and longitude, 93–94
    log-log, 98–99
    navigator's, 87–89
    polar, 80–92
    semilog, 96–97
    spherical, 110–113
    time-space, 115–118
cosecant
    function, 169, 232
    of negative angle, 241

cosine
    function, 84–85, 230–231
    of angular difference, 242
    of angular sum, 242
    of double angle, 242
    of negative angle, 240
    Pythagorean Theorem involving, 238
    wave, 323
cotangent
    function, 233
    of negative angle, 241
counting numbers, 7
cross-multiplication, 38
cross product
    of mixed vectors and scalars, 276
    of two vectors, 253–254, 258, 262–263, 272
cube
    definition of, 188
    surface area of, 188
    volume of, 188
cubic equation, 134–137
cylinder
    definition of, 192
    right circular, 195–196
    slant circular, 196–197
cylindrical coordinates, 109–110, 113

D
decagon, 179
decimal
    expansion, 9
    form, 9, 55
    number system, 10, 12–13
    point, 9, 52
declination
    in celestial coordinates, 94–95, 99
    in compass direction, 88
    in spherical coordinates, 110–111
definite integral, 339–341
degree of arc, 228–229
DeMorgan's law
    for conjunction, 219
    for disjunction, 219
denumerable number set, 6–10

dependent variable, 73, 107, 306
derivative
    chain rule for, 313
    definition of, 305–311
    first, 308
    higher-order, 309
    notation used for, 309
    of constant function, 311
    of function multiplied by constant, 312
    of difference of two functions, 312
    of down-ramp wave, 325
    of product of functions, 312
    of quotient of two functions, 313
    of sum of functions, 311
    of variable raised to a power, 313
    of sine wave, 323–324
    of square wave, 326–327
    of triangular wave, 325–327
    of up-ramp wave, 324–325
    of wave function, 323–327
    reciprocal, 313
    second, 309
    table, 328
difference
    of exponents, 41
    of functions, 311
    powers of, 41
differential
    calculus, 305–335
    in expression of integral, 338
differentiation, 305–335
direction
    angles, 261–262
    cosines, 262
    in polar coordinates, 82
    of vector, 22, 252–253
discriminant, in quadratic equation, 132
disjoint sets, 4
disjunction, logical, 209–210, 213, 219
displacement, 354
distributive law
    of conjunction over disjunction, 219
    of cross product over vector addition, 276

of division over addition, 38
of dot product over vector addition, 275–276
of multiplication over addition, 32
of scalar multiplication over scalar addition, 275
of scalar multiplication over vector addition, 275
division
    and significant figures, 61
    by zero, 34
    using scientific notation, 59–61
domain of function, 285, 307
dot product
    of two vectors, 253, 258, 262, 272
    of cross products, 276
    of mixed vectors and scalars, 276
double angle, 242
double negation, logical, 217
down-ramp wave
    derivative of, 325–326, 327, 329
    integral of, 350–351

E

element of set, 1
elevation, in spherical coordinates, 112
ellipse
    axes of, 174
    definition of, 174
    interior area of, 174–175
    semi-axes of, 85, 174
ellipsoid
    definition of, 199
    semi-axes of, 199–200
    volume of, 199–200
equality symbol, approximate, 62
equation
    2-by-2 general, 152–154
    2-by-2 linear, 143–148
    3-by-3 linear, 148–152
    cubic, 134–137
    graphic solution of, 154–158
    higher-order, 134–137
    linear, 74–80, 127–130, 143–152
    nonlinear, 152–154
    nth-order, 136

quadratic, 130–134
quartic, 135
quintic, 135
multivariable, 143–162
single-variable, 125–141
equivalence, logical, 211–212, 215
Euler's constant, 283
exclusive OR, 209–210
exponent
  negative, 40
  plain-text, in scientific notation, 53–54
  rational-number, 40
exponential
  alternative expressions for, 292
  base-10, 290–291
  base-e, 290–293
  behavior of, 293–298
  common, 290–291
  common and natural, product of, 295
  common to natural, ratio of, 295
  constant, 283
  definition of, 283
  function, 290–298
  natural, 290–293
  natural to common, ratio of, 295
  of exponential, 294
  of ratio, 294
  product of, 294
  reciprocal of, 294
  versus logarithm, 293–294
exponentiation
  and significant figures, 61
  in scientific notation, 59, 61

F
facet of polyhedron, 185
factorial, 35
first derivative, 308
frustum of cone, 193–194
function
  average value over interval, 356–357
  constant, 311
  continuous, 308

definition of, 306–307
domain of, 285, 307
inflection point of, 320–321
local maximum of, 318–319
local minimum of, 318–319
multiplied by constant, 312
range of, 285, 307
single-variable, 306
zero, 311
Fundamental Theorem of Calculus, 339–340

G
geographic north, 87, 109–110
geometric mean, 35
Global Positioning System (GPS), 88

H
heading, 87
height, in cylindrical coordinates, 109
hexadecimal number system, 11–15
higher-order
  derivative, 309
  equation, 134–137
hour, in right ascension, 94–95
hyperspace, 113–119
hypotenuse, 22–23, 164–165

I
identities, trigonometric, 237–245
identity element
  additive, 30
  multiplicative, 30
IF/THEN operation, 211
imaginary number, 17–18
impedance, 22, 35
implication, logical, 210–211, 213–214
implication reversal, 218–219
inclusive OR, 209–210
indefinite integral
  definition of, 338
  of constant, 341
  of constant raised to variable power, 345
  of function multiplied by constant, 342

indefinite integral, *continued*
  of reciprocal, 344
  of reciprocal multiplied by constant, 344–345
  of sum of functions, 343
  of variable, 342
  of variable multiplied by constant, 342
  of variable raised to integer power, 342
  table, 346
independent variable, 73, 107, 306
"infinity," 20
inflection point of function, 320–321
instantaneous rate of change, 305
integral
  calculus, 337–364
  definite, 339–341
  indefinite. *See* indefinite integral
  of down-ramp wave, 350–351
  of sine wave, 348–350
  of square wave, 351–352
  of triangular wave, 351–352
  of up-ramp wave, 350
  of wave function, 348–354
integration
  by parts, 344
  constant of, 338–339
  definition of, 337–341
integrator, 348
interior angle
  of quadrilateral, 166
  of triangle, 163–164
interior area
  of circle, 173–175
  of circular sector, 178
  of ellipse, 174–175
  of parallelogram, 166–167
  of quadrilateral, 166–171
  of rectangle, 169
  of regular polygon, 171–172, 175–177
  of rhombus, 167
  of square, 168–169
  of trapezoid, 169–170
  of triangle, 164
intersection of sets, 3–4
inverse

additive, 30
  logarithm, 292
  multiplicative, 30–31
irrational number, 15–16

J
j operator, 17

L
latitude
  celestial, 94–95, 98–99, 112–113
  terrestrial, 93–94
line
  normal to curve, 316–317
  slope of, 75, 157–158
  tangent to curve, 308, 315–316
lines, parallel, 158
linear equation
  2-by-2, 143–148
  3-by-3, 148–152
  finding, based on graph, 79–80
  graph of, 155–158
  point-slope form of, 78–79
  single-variable, 127–130
  slope-intercept form of, 75–77, 156–157
  standard form of, 74–75
linear scale, 55
linearity, principle of, 344, 345, 346
local
  maximum of function, 318–319
  minimum of function, 318–319
logarithm
  base-10, 284–285
  base-e, 285–286
  behavior of, 287–290
  common, 284–285
  common in terms of natural, 288
  definition of, 283
  Napierian, 285–286
  natural, 285–286
  natural in terms of common, 288
  of base raised to power, 288
  of power, 287
  of product, 287

of ratio, 287
of reciprocal, 287
of root, 288
versus exponential, 293–294
logarithmic
    function, 284–290
    scale, 54–55, 96–99
log-log coordinates, 98–99
logical
    conjunction, 208–209, 213
    contradiction, 217
    contrapositive, 218–219
    disjunction, 209–210, 213
    double negation, 217
    equivalence, 211–212, 215
    implication, 210–211, 213–214
    implication reversal, 218–219
    negation, 208, 212–213
longitude
    celestial, 94–95, 98–99, 112–113
    terrestrial, 93–94

M

magnetic north, 88
magnitude of vector, 22, 252–253
major axis of ellipse, 174
major semi-axis of ellipse, 85, 174
mathematical symbols, 2
mathematician's polar coordinates, 80–82
maximum of function, local, 318–319
mean
    arithmetic, 35
    geometric, 35
member of set, 1
minimum of function, local, 318–319
minor axis of ellipse, 174
minor semi-axis of ellipse, 85, 174
minute
    of arc, 95, 228–229
    of right ascension, 94–95
modulo-2 number system, 11–13
modulo-8 number system, 11–13
modulo-10 number system, 10, 12–13
modulo-16 number system, 12–13

multiplication
    and significant figures, 61
    associative law of, 32
    by zero, 34
    commutative law of, 31
    of vector and scalar, 272, 274
    using scientific notation, 57–58, 60–61
multiplicative
    identity element, 30
    inverse, 30–31, 287
multivariable equation
    2-by-2 general, 152–154
    2-by-2 linear, 143–148
    3-by-3 linear, 148–152
    graphic solution of, 154–158
    inconsistent, 148

N

Napierian logarithm, 285–286
natural exponential
    definition of, 290–293
natural logarithm
    definition of, 285–286
    in terms of common logarithm, 288
natural number, 7
navigator's coordinates, 87–89
negation, 208, 212–213
negative
    angle, 240–241
    exponent, 40
    number, 9
nondenumerable number set, 15–20
nonstandard angle, 233–234
nonterminating, repeating decimal number, 9
North Star, 88
NOT operation, 208
n-space, Cartesian, 118–119
nth-order equation, 136
number
    binary, 11–13
    complex, 18–24
    counting, 7
    decimal, 10, 12–13
    hexadecimal, 11–13

number, *continued*
imaginary, 17–18
integers, 7–8
irrational, 15–16
line, 54–55
natural, 7
negative, 9
nonterminating, repeating decimal, 9
octal, 11–13
rational, 8–9
real, 16
terminating decimal, 9
transfinite cardinal, 16–17
whole, 7

O
obtuse angle, 228
octagon, 172
octal number system, 11–13
one-variable equation. *See* single-variable equation
OR operation, 209–210
order of magnitude, 54–55, 297
ordered
pair, 72–73
quadruple, 114
triple, 106–107
ordinate, 73, 229
origin
in Cartesian plane, 73
in polar coordinates, 80
originating point of vector, 254, 264–265

P
parabola, 155
parallel lines, 158
parallelepiped
definition of, 189–190
surface area of, 189–190
volume of, 190
parallelogram
definition of, 166
interior area of, 166–167
perimeter of, 166–167
perimeter
of circular sector, 177–178

of parallelogram, 166–167
of quadrilateral, 166–171
of rectangle, 180
of regular polygon, 171–172, 175–177
of rhombus, 167
of square, 168–169
of trapezoid, 169–170
of triangle, 164
point-slope form of linear equation, 78–79
polar coordinates
conversions to and from, 89–92
direction in, 82
functions rendered in, 310
graphs in, 83–88
mathematician's, 80–82
origin in, 80
radius in, 80–82
vectors in, 256–259
Polaris, 88
polygon, regular. *See* regular polygon
polyhedron
definition of, 185
facet of, 185
power
logarithm of, 287
of 10, 51–57
of signs, 36
precedence
in calculations, 44–45
in Boolean expressions, 216
prefix multipliers, 56
primary circular function, 229–232
principle of linearity, 344, 345, 346
prism
rectangular, as room interior, 190–191
rectangular, definition of, 188
rectangular, surface area of, 189
rectangular, volume of, 198
product
logarithm of, 287
of exponentials, 294–195
of exponents, 41
of functions, 312
of quotients, 39
of signs, 36

of sums, 37–38
reciprocal of, 38
proper subset, 5
pyramid
    rectangular, definition of, 187
    rectangular, volume of, 187
    symmetrical square, definition of, 187
    symmetrical square, surface area of, 187
Pythagorean Theorem
    for right triangle, 22–23, 73, 164–165, 236–240
    for secant and tangent, 238
    for sine and cosine, 238

Q

quadratic
    equation, 130–134, 155–156
    formula, 131–132
quadrilateral
    definition of, 166
    interior angle of, 166
    interior area of, 166–171
    perimeter of, 166–171
    side of, 166
    vertex of, 166
quartic equation, 135
quintic equation, 135
quotient
    of exponents, 41
    of functions, 313
    of quotients, 39
    of signs, 36
    reciprocal of, 39

R

radian, 228–229
radius
    in cylindrical coordinates, 109
    in polar coordinates, 80–82
    in spherical coordinates, 110–112
radix-2 number system, 11–13
radix-8 number system, 11–13
radix-10 number system, 10, 12–13
radix-16 number system, 12–13
radix point, 9, 52–53
range

in navigator's coordinates, 87
    of function, 285, 307
rate of change, instantaneous, 305
ratio
    exponential of, 294
    logarithm of, 287
    of exponentials, 294–295
rational number
    definition of, 8–9
    in exponent, 40
real number, 16
real-number coefficient, 17
reciprocal
    definition of, 36
    derivative, 313
    logarithm of, 287
    of exponential, 294
    of product, 38
    of quotient, 39
    of reciprocal, 36–37
rectangle
    definition of, 168
    interior area of, 168
    perimeter of, 168
rectangular
    3D coordinates, 105–105
    coordinate plane, 71–72
    prism, 188–189
    pyramid, 187
reference axis, 251
regular polygon
    circumscribing circle, 176–177
    definition of, 171
    inscribed in circle, 175–176
    interior area of, 171–172, 175–177
    perimeter of, 171–172, 175–177
relation, 306–307
rhombus
    definition of, 167
    interior area of, 167
    perimeter of, 167
right ascension, 94–95, 99, 110–111
right
    angle, 164–165
    triangle, 164–165, 234–237

right-hand rule for cross product of vectors, 254
root
    logarithm of, 288
    positive integer, 34–35
    using scientific notation, 59
rounding, 43–44

S
scalar product of vectors, 253
scientific notation
    addition using, 59–60, 64–65
    alternative form of, 52
    calculations in, 57–61
    coefficient in, 52
    division using, 58, 60–61
    exponentiation in, 59, 61
    multiplication using, 57–58, 60–61
    multiplication symbol in, 52–53
    orders of magnitude in, 54–55
    plain-text exponents in, 53–54
    powers of 10 in, 51–57
    roots using, 59
    standard form of, 51–52, 57
    subtraction using, 60, 64–65
    when to use, 55–56
secant
    function, 232–233
    of negative angle, 241
    Pythagorean Theorem involving, 238
second
    of arc, 95, 228–229
    of right ascension, 94–95
second derivative, 309
secondary circular function, 232–234
semi-axes
    of ellipse, 85, 174
    of ellipsoid, 199–200
semilog coordinates, 96–97
sense of concavity, 320
set
    cardinality of, 5
    definition of, 1–3
    denumerable, 6–10
    element of, 1

intersection, 3–4
member of, 1
nondenumerable, 15–20
union, 3–5
sets
    coincident, 3
    disjoint, 4
side
    adjacent, 236
    of quadrilateral, 166
    of triangle, 164
significant figures
    addition and, 64–65
    accuracy and, 62
    division and, 61
    exact values and, 63
    exponentiation and, 61
    multiplication and, 61
    subtraction and, 64–65
    zeros as, 62
signs
    power of, 36
    product of, 36
    quotient of, 36
sine
    function, 85, 229–230
    of angular difference, 242
    of angular sum, 242
    of double angle, 242
    of negative angle, 240
    Pythagorean Theorem involving, 238
    wave, derivative of, 323–324
    wave, integral of, 348–350
single-variable equation
    linear, 127–130
    quadratic, 130–134
    operational rules for, 125–127
slope of line, 75, 157–158
slope-intercept form of linear equation, 75–77
sphere
    definition of, 198
    surface area of, 198–199
    volume of, 198–199
spherical coordinates, 110–113

speed, 354–358
spiral of Archimedes, 85–86
square
    definition of, 168
    interior area of, 168–169
    perimeter of, 168–169
square wave
    derivative of, 325–327
    integral of, 351–352
standard form of linear equation, 74–75
substitution method
    for solving 2x2 linear equations, 143–145,
        146–147
subtraction
    and significant figures, 64–65
    using scientific notation, 60, 64–65
sum
    of exponents, 40
    of functions, 311
    of quotients, 40
    of vectors, 253, 257, 260–261
    powers of, 41
surface area
    of cube, 188
    of frustum of cone, 193–194
    of parallelepiped, 189–190
    of rectangular prism, 189
    of right circular cone, 193
    of right circular cylinder, 196
    of sphere, 198–199
    of symmetrical square pyramid, 197
    of tetrahedron, 186
    of torus, 200–201
symbols, mathematical, 2

T

tangent
    function, 230–232
    of negative angle, 240–241
    Pythagorean Theorem involving, 238
terminating decimal number, 9
tetrahedron
    definition of, 186
    surface area of, 186

volume of, 186
Theorem of Pythagoras
    for right triangle, 22–23, 73, 164–165,
        236–240
    for secant and tangent, 238
    for sine and cosine, 238
time-space, 115–118
torus
    definition of, 200
    surface area of, 200–201
    volume of, 200–201
transfinite cardinal number, 16–17
trapezoid
    definition of, 169
    interior area of, 169–170
    perimeter of, 169–170
triangle
    definition of, 163
    interior angle of, 163–164
    interior area of, 164
    notation for, 235
    perimeter of, 164
    right, 164–165, 234–237
    side of, 163
    sum of angle measures in, 236
    vertex of, 163
triangular wave
    derivative of, 325–327
    integral of, 351–352
trigonometry, 227–250
trigonometric identities, 237–245
truncation, 43
truth table
    for conjunction, 213
    for disjunction, 213
    for implication, 213–214
    for logical equivalence, 215
    for negation, 212–213

U
union of sets, 3–5
unit circle, 227–229
unit imaginary number, 17
unit vector, 270–272

up-ramp wave
    derivative of, 324–325, 327, 329
    integral of, 350

V
variable
    dependent, 73, 107, 306
    independent, 73, 107, 306
vector
    addition, 272–274
    back-end point of, 254
    component, 270–271
    cross product, 253–254, 258, 262–263, 272–273, 276
    definition of, 22–23
    direction of, 22, 252–253, 261–262
    dot product, 253, 258, 262, 272, 276
    equivalent, 265–266
    identical, 265
    in Cartesian plane, 251–256
    in Cartesian 3-space, 259–264
    in polar plane, 256–259
    magnitude of, 22–23, 252, 261
    multiplication by scalar, 267–270, 272, 274
    originating point of, 254, 264–265
    portrayal of, 251
    product, 253–254
    right-hand rule for cross product, 254
    scalar product, 253
    standard form of, 264–267
    sum, 253, 257, 260–261
    unit, 270–272
vertex
    of quadrilateral, 166
    of triangle, 163
volume
    of cube, 188
    of ellipsoid, 199–200
    of frustum of cone, 194
    of parallelepiped, 193
    of rectangular prism, 189
    of rectangular pyramid, 187
    of right circular cone, 193
    of right circular cylinder, 196
    of slant circular cone, 195–195
    of slant circular cylinder, 197
    of sphere, 198–199
    of tetrahedron, 186
    of torus, 200–201

W
wave function
    derivative of, 323–327
    integral of, 348–354
whole numbers, 7

X
x axis in Cartesian plane, 72
xyz-space, 106

Y
y axis in Cartesian plane, 72
y-intercept, 76, 157–158

Z
zero
    as a significant figure, 62
    denominator, 34
    function, 311
    multiplication by, 34
    numerator, 33
zeroth power, 34

ABOUT THE AUTHOR

Stan Gibilisco is one of McGraw-Hill's most prolific and popular authors. His clear, reader-friendly writing style makes his books accessible to a wide audience, and his experience as an electronics engineer, researcher, and mathematician makes him an ideal editor for reference books and tutorials. He has authored several titles for the McGraw-Hill *DeMYSTiFied* series (a group of home-schooling and self-teaching volumes), including *Everyday Math Demystified*, *Physics Demystified*, and *Statistics Demystified*, all perennial bestsellers. Stan has also written more than 20 other books and dozens of magazine articles. His work has been published in several languages. *Booklist* named his *McGraw-Hill Encyclopedia of Personal Computing* one of the "Best References of 1996," and named his *Encyclopedia of Electronics* one of the "Best References of the 1980s."